CW01461185

ARE VIRUSES
ALIVE?

ARE VIRUSES ALIVE?

Mind-Altering Stories about Life and Evolution

NOGA WIES

PELAGIC PUBLISHING

First published in 2025 by
Pelagic Publishing
20–22 Wenlock Road
London N1 7GU

www.pelagicpublishing.com

Are Viruses Alive? Mind-Altering Stories about Life and Evolution

Copyright © 2025 Noga Wies

The moral rights of the author have been asserted by her
in accordance with the Copyright, Designs and Patents Act 1988.

All rights reserved. Apart from short excerpts for use in research or for
reviews, no part of this document may be printed or reproduced, stored in
a retrieval system, or transmitted in any form or by any means, electronic,
mechanical, photocopying, recording, now known or hereafter invented or
otherwise, without prior permission from the publisher.

https://doi.org/10.53061/NXIH8148

A CIP record for this book is available from the British Library

ISBN 978-1-78427-576-1 Hbk
ISBN 978-1-78427-577-8 Pbk
ISBN 978-1-78427-578-5 ePub
ISBN 978-1-78427-579-2 PDF

EU Authorised Representative:
Easy Access System Europe – Mustamäe tee 50, 10621 Tallinn, Estonia,
gpsr.requests@easproject.com

Cover image: Kateryna Kon / Science Photo Library

Typeset in Minion Pro by S4Carlisle Publishing Services, Chennai, India

For Daniel and Dawn

Contents

Preface

There is no greater thrill than the rush of discovering a game-changing idea – the kind that turns your perception of reality on its head. Take, for example, artificial intelligence. You spend your whole life assuming that robots will gradually introduce objectivity to the courts, our educational system and everything in between, only to discover that robots regurgitate the biases of the data they are trained on, perpetuating the ugliest of human prejudices. When it comes to economic issues, you are taught that communism, with its made-up jobs and surplus production, is the epitome of economic inefficiency, only to learn that capitalistic bureaucracy can give it a run for its money. The best game-changing ideas are so shocking that nothing is ever the same after absorbing them.

When it comes to what we know about life, our perspective often tends to be downright boring. As toddlers, we are taught about giraffes with long necks, elephants with long trunks, and itsy-bitsy spiders with eight legs. This effectively indoctrinates us to think of living things as discrete entities characterized by salient distinguishing features. Later, in middle school, we learn that living things are made up of cells. And some of us then go on to gain a basic understanding of natural selection in high school or university. This results in our viewing ourselves as distinct, multicellular creatures that can, say, become incrementally furrier in response to a cold climate. But this is a terribly simplistic view of what we are. There is so much more to the fascinating and freaky phenomenon we call life. In fact, our existence can be interpreted from (at least) ten game-changing scientific perspectives! Curious to know the real meaning(s) of life? Read on to see how deep the rabbit hole goes.

CHAPTER 1

Genial Genes

Our story begins with one of the most powerful forces in the universe – British classism. In 1831 Captain Robert FitzRoy of the Royal Navy was appointed commander of the HMS *Beagle* and tasked with surveying the South American coast. The voyage was likely to be a long one, and Captain FitzRoy could hardly have been expected to stoop to dining with his inferiors. Unwilling to spend many lonely years at sea, Captain FitzRoy searched for a gentleman companion who could dine with him as his equal. John Stevens Henslow, a professor at Cambridge, recommended a young naturalist named Charles Robert Darwin for the post, and the rest, as they say, is history.

Darwin's unquenchable curiosity and debilitating seasickness led him to spend much of the *Beagle*'s nearly five-year voyage on land, collecting countless plants, animals and fossils. While exploring the Galapagos Islands, he collected birds from various islands that were mistakenly believed to be finches. Darwin's finches, as they are now known, were nearly identical save for great differences in beak size and shape. Why did finches from different islands possess different kinds of beaks? Darwin was familiar with his grandfather Erasmus Darwin's idea that species change over time, and therefore readily concluded that his specimens shared a common ancestor and had developed distinct beaks over many generations. But did they have enough time to do so? Fortunately for Darwin, Captain FitzRoy had given him a copy of Charles Lyell's book *Principles of Geology*, which posited an ancient Earth with geological features forming gradually over eons. Lyell's theory provided Darwin's finches with plenty of time to evolve a variety of beaks.

Back in England, Darwin stumbled upon Thomas Robert Malthus's book *An Essay on the Principle of Population*, which argued that with the human population growing at a rate far exceeding that of the increase in food production, famine was inevitable. When Darwin applied this book's ideas to nature, it dawned upon him that plants and animals compete for limited resources, and that consequently, only some of their offspring survive to adulthood. But what, exactly, determines which of the offspring survive and reproduce? Having grown up on a country estate, Darwin was familiar with the agricultural practice of selective breeding. Selective breeding involves the selection of plants or animals with desirable traits to breed together, thus producing offspring with increasingly desirable traits. Since human farmers were intervening to choose which individuals were allowed to breed, selective breeding could be termed *artificial selection*. Nature, Darwin realized, was like an all-encompassing farmer who constantly selects for traits that are conducive to survival, in a process he dubbed *natural selection*.

Let us return to Darwin's finches. Long before Darwin visited the Galapagos, a bird from the mainland produced offspring with identical beaks. These offspring flew away and settled on various islands, where they encountered different conditions. Say, for example, that large nuts were common on one of the Galapagos Islands. Finches who happened to settle on this island would benefit from large beaks, as these would enable them to eat more nuts, thus increasing their chances of survival and reproduction. That is why finches who happened to possess slightly larger than average beaks produced more offspring, who inherited their parents' larger beaks. This is how natural selection works. Larger beaks were selected for over and over again, until the island's finches had beaks that were substantially larger than their mainland ancestor's beak. Meanwhile, the beaks of finches on other islands were gradually evolving to adapt to a wide range of food sources, resulting in

Nonsense!!

No!

X

the great variety of beaks observed by Darwin. Darwin's theory of evolution by natural selection, detailed in his revolutionary 1859 book *On the Origin of Species*, explained how natural selection drove incremental changes to traits such as beak size, which accumulated over many generations to produce new species.

Not Darwin

Darwin's elegant theory is all the more impressive when one considers that Darwin formulated it with no idea of how heredity works. Think about it – finches don't just inherit their parents' beak size, they also inherit the shade of their feathers, the pitch of their song, etc. And their ability to inherit all these traits is integral to evolution. But how do they do it? Darwin sought to fill this gap in his theory by coming up with a speculative hypothesis known as 'pangenesis'. He was wary of publishing an idea as yet unsupported by evidence, and his friend Thomas Huxley agreed that this would be a bad idea – at first. However, Huxley ultimately changed his mind and wrote to Darwin that 'somebody rummaging among your papers half a century hence will find pangenesis and say "see this wonderful anticipation of our modern theories – and that stupid ass, Huxley, prevented his publishing them".' Darwin was convinced, and proceeded to publish an explanation of pangenesis in 1868. According to this hypothesis, organs constantly emit tiny organ-specific particles called gemmules. The gemmules possess 'mutual affinity' which causes them to accumulate in the gonads, or ovaries and testes. Once they are passed on to offspring, each gemmule develops into an organ that expresses the traits of the gemmule's organ of origin. Despite Darwin's efforts, pangenesis never caught on, which is fortunate because it turned out to be utterly wrong.

* * *

Darwin's pangenesis mechanism can be interpreted to imply *blending inheritance*, a hypothesis accepted by many nineteenth-century biologists. According to this hypothesis, parental traits blend together in offspring, meaning that every one of your traits strikes an average between the correspond-ing traits of your parents. This is patently untrue; and when you think about it, if it *were* true, it would result in individuals becoming more and more alike with every generation, until no trace of variation remained in the population. But if pangenesis and blending inheritance are wrong, how *does* heredity work? The key to this mystery lies with an inquisitive, and perhaps calculating, monk.

Gregor Johann Mendel was an Austrian who, partly motivated by his inability to personally finance his studies, joined a monastery in 1843. A keen botanist, he conducted groundbreaking research on pea plants that he bred in the monastery's garden, seeking to shed light on the heredity of traits such as height, flower position, pea shape and color. To understand why Mendel's ideas were so revolutionary, let's look at pea color. Mendel grew pure-bred plants with green peas and pure-bred plants with yellow peas. According to the blending inheritance hypothesis, cross-breeding green pea plants with yellow pea plants should produce hybrid offspring with some in-termediate pea color. But when Mendel crossed his plants, all the offspring had yellow peas! Mendel dubbed the yellow pea color a 'dominant' trait because it was expressed in all the hybrid plants. He then crossed the hybrid plants, and was surprised to see the re-emergence of the green pea trait in a quarter of their offspring. Mendel dubbed the green pea color a 'recessive' trait, because it had been suppressed by the dominant trait.

To make sense of his discoveries, Mendel proposed that when two plants are crossed, each one passes down a distinct factor which contains information regarding pea color. The original pure-bred plants had either two green factors or two yellow factors, so when they were crossed, their offspring all had one green factor and one yellow factor. This yielded yellow peas because the yellow factor is dominant, while the recessive green factor failed to express itself. When the offspring were crossed, three quarters of the resulting plants produced yellow peas because half inherited one green and one yellow factor, and an additional quarter inherited two yellow factors. However, a quarter of the plants ended up with two green factors and consequently produced green peas.

Mendel's theory is known as *particulate inheritance* because unlike *blending inheritance*, it posits the existence of discrete hereditary particles that are passed down separately from generation to generation. This allows for the preservation of variation in populations, a key component of Darwinian natural selection. Sadly, Mendel, now considered the father of modern genetics, did not live to see his theory win over the scientific community. He presented his ideas at two meetings of the Natural History Society of Brno in 1865 and published a paper in 1866,[2] but his experiments failed to garner attention before his death in 1884. It was only in 1900 that three researchers from three different countries independently rediscovered Mendel's work, propelling particulate inheritance to a position of scientific acclaim.

When statisticians later analyzed Mendel's experimental results, they discovered that they were, well, too good to be true.[3] Think of it this way: if you flip a coin several times, you expect to get heads around 50% of the time and tails around 50% of the time. But if someone reported flipping a coin ten times a day for a week and always getting heads five times and tails five times, you'd be suspicious. Similarly, Mendel appears to have obtained percentages that fit his theory almost perfectly due to confirmation bias.* Still, his statistical sins did little to detract from his monumental contribution to science.

* * *

Once the theory of particulate inheritance was proposed by a monk, it was time for a man who had nearly become a priest to characterize the hereditary particles themselves. Growing up,

---

* Confirmation bias is the human tendency to seek out or focus on information that confirms previously existing beliefs.

the Swiss physician Johannes Friedrich Miescher dreamed of becoming a priest, but his father disapproved, and he ended up in med school instead. When he graduated in 1868, Miescher realized that his partial deafness would make it difficult to interact with patients and decided to focus on research. He began studying the nucleus, the largest of the cell's organelles, and extracted white blood cell nuclei from pus found on used bandages he received from a local hospital. A talented researcher, Miescher was the first to isolate a molecule he dubbed *nuclein*,[4] which is now known as DNA. And though he never proved it, Miescher speculated that DNA was involved in heredity, perhaps by encoding a hereditary alphabet of some sort.[5]

Miescher's ideas were gradually validated when a German biochemist named Albrecht Kossel isolated the four building blocks, called *nucleotides*, of DNA between 1885 and 1901. Kossel, who viewed the subjects of his research as actors in the drama of life, discovered the nucleotides adenine (A), cytosine (C), guanine (G), and thymine (T),[6] and was rewarded with the 1910 Nobel Prize in Physiology or Medicine. Still, Kossel only proved that a four-letter hereditary alphabet *could* exist, not that it *did* exist. This next accomplishment fell to two legendary teams.

The first of these consisted of Oswald Avery, Colin MacLeod and Maclyn McCarty. Before their famous 1944 experiment, the British bacteriologist Frederick Griffith had demonstrated that hereditary traits could be transferred between bacteria. Griffith worked with two strains of pneumococcus bacteria – a virulent strain, which killed the mice it infected, and a non-virulent strain. In a landmark experiment known today simply as 'Griffith's experiment', he killed the virulent bacteria and added their remains to the non-virulent bacteria before injecting them into mice. The mice died, and Griffith reasoned that hereditary particles from the dead bacteria had found their way into the live bacteria, effectively transforming them into virulent bacteria.[7] But what, exactly, were these hereditary particles?

When Avery, MacLeod and McCarty set out to answer this question, proteins were widely believed to be responsible for heredity. However, when the researchers repeated Griffith's experiment under protein-degrading conditions, the mice still died, ruling out proteins as the carriers of hereditary information. Avery, MacLeod and McCarty eventually isolated the hereditary particle and performed biochemical tests to elucidate its composition. Much to their surprise, it turned out to be DNA[8] – a molecule then commonly assumed to serve a strictly structural purpose.

Now that DNA had been singled out as the hereditary particle, a second team of scientists set out to decode the language of heredity. Francis Crick, an English molecular biologist who won the 1962 Nobel Prize in Physiology or Medicine for deciphering the double-helical structure of DNA, had always been fascinated by how inanimate molecules create life. He suspected that the sequence of nucleotides in DNA encoded heredity traits, and teamed up with Sydney Brenner, Leslie Barnett and Richard Watts-Tobin to crack the code. In 1961, Crick and his colleagues conducted a strikingly elegant experiment in which they added or removed nucleotides from a virus's DNA. Adding or removing one or two nucleotides was shown to impair functionality, whereas adding or removing three nucleotides maintained functionality. This, the researchers deduced, could only mean that hereditary information is encoded in triplets of DNA nucleotides.[9] Put simply, if nucleotides are letters in the language of DNA, then triplets of nucleotides are the words that record our hereditary traits.

* * *

But what do these words mean? And how do they determine everything about us? Before we dive into the answer to these all-important questions, let us first take a moment to appreciate proteins. There are many, many different kinds of proteins. Humans, for example, possess roughly 20,000 different proteins, while the *E. coli* bacteria in our gut possess over 4,000.[10] This astonishing variety of proteins translates into a rich tapestry of protein functions. Some proteins are involved in signaling processes, either as the signaling molecules themselves or as the receptors they bind to pass information. Others are structural proteins that make up countless components of our cells and their surroundings. Contractile proteins enable muscle function, while plasma proteins do everything from maintaining blood pressure to fighting infections. And protein enzymes are minuscule biological machines that speed up the vital chemical reactions occurring within us. Proteins, in short, do everything in living creatures and, consequently, determine their biological traits.

The awe-inspiring diversity of proteins stems from their molecular structure. Proteins are long chains of molecules known as amino acids, and the precise sequence of amino acids determines a protein's function. But how do proteins get their specific sequences? This is where DNA comes in. Each triplet of DNA nucleotides, or *codon*, is translated into a specific amino acid, thus transferring the hereditary information stored in DNA to protein sequences. By discovering codons, Crick and his colleagues paved the way for the deciphering of each codon's amino acid 'meaning', a process that was finally completed in 1966.

Let us return to Darwin's theory of evolution by natural selection. We now know that each protein is encoded by a stretch of DNA known as a gene. Genes are the hereditary particles that fill in the gap in Darwin's theory, which is often summarized as 'survival of the fittest'. This term is so well known that few of us stop to ask – the fittest what, exactly? Is it the fittest individual? This sounds plausible, as individual organisms are constantly battling it out in the struggle for life. As we've discussed, a finch who happens to have a gene for a larger beak will likely thrive on an island with large nuts, while a finch with a gene for a smaller beak may well starve before reaching adulthood. But life isn't all about an individual's fitness. Consider, for instance, that meerkats volunteer for guard duty,[11] choosing to spend time they could dedicate to their own survival and reproduction protecting others as they search for food. Their actions seem to indicate that evolution is really about the survival of the fittest *group*.

Group fitness is exemplified by ants. Ants are eusocial, which means that both reproductive and non-reproductive castes live together in highly organized colonies where individuals of all ages cooperate seamlessly to rear communal young. The ants in a colony are so in sync, in fact, that the colony itself may be considered a *superorganism*. Moreover, since an ant colony will often integrate information obtained by individual ants to solve extremely complex problems, it can also be described as a 'super mind'.

A typical ant colony is home to four castes. The first, and most populous, consists of workers. Workers are females who are incapable of reproducing. They build and maintain the colony, defend it from rival colonies, forage, and feed the young. Queens, on the other hand, do nothing *except* reproduce, often laying hundreds of eggs a day. Next up are drones, the male ants who mate with the queen, and then, having served their purpose, drop dead. Last but not least, reproductive ants are winged individuals who leave their birth colony to establish new colonies elsewhere.

Given the advanced specialization and seamless cooperation found in ant colonies, it is surprising to consider how readily both workers and queens may betray their colonies. Some new ant colonies are polygynous, meaning that they have multiple queens. Once the first generation of workers hatch, they wage both conventional and chemical warfare against the queens and sometimes kill all of them, thus dooming the entire colony. As for the queens, if they sense that their lives may be in danger due to the presence of other queens, they cheat the colony by laying fewer eggs in an effort to conserve energy for self-defense.[12]

Perhaps, then, it is the fittest *species* that survive. This idea is supported by the geographic isolation of the cuddliest creature of all – the koala. Koalas are marsupials, which means that their females have a *marsupium*, or pouch, for carrying their offspring. But why must marsupials carry their young in a pouch? The answer is that marsupials have short gestation times, resulting in the birth of under-developed young who spend weeks or months completing their development in their mother's pouch. Placental mammals, on the other hand, develop a placenta – an organ that attaches to the uterine wall and provides the developing fetus with plenty of oxygen and nutrients from the mother's blood. This extends gestation times and guarantees the birth of fully developed young.

Marsupials such as koalas, kangaroos and wombats are well-known Australian species, so it may come as a surprise that marsupials actually originated in North America around 125 million years ago. Sixty or so million years later, they made their way down to South America by island-hopping, as the Panama land bridge did not yet link North America and South America. What is now known as the continent of South America was connected to a lush Antarctica at the time, which, in turn, was connected to Australia. Ancient marsupial fossils discovered on an Antarctic island prove that marsupials crossed into Antarctica before completing their epic journey to Australia.[13] There they diversified into everything from giant kangaroos and wombats to *Thylacoleo carnifex*, the 'meat-cutting marsupial lion' who roamed the once densely forested Australian wilderness.

Back in South America, many marsupials went extinct when the continent was finally linked to North America, as this allowed placental mammals to cross into South America and outcompete the native marsupials with more developed, and therefore less vulnerable, young. Meanwhile, Australia's new-found geographic isolation prevented the introduction of placental competitors, enabling sustained marsupial dominance and diversity. In other words, a species' success or extinction depend on its biological traits. This lends credence to the idea that evolutionary forces act on species, not groups.

* * *

But what is a species, exactly? Thomas Samuel Kuhn, a renowned philosopher of science, famously bemoaned that scientific research entails the forcing of nature into conceptual boxes.[14] This holds true for the conceptual boxes we call species, which, while useful for classifying certain organisms, break down completely when we attempt to classify others. Take, for example, the organisms inhabiting a house. Classifying them as humans, dogs, cats, goldfish, bedbugs and so on is a simple way to make sense of reality. But organisms don't always fit into discrete boxes so neatly. To understand why, we must first define what a species is.

According to the *Encyclopedia Britannica*, a species is comprised of related organisms that share common characteristics and are capable of interbreeding. Using this definition to classify our domestic organisms is easy, as the human inhabitants of a house differ greatly from its bedbugs and can hardly interbreed with them. But what about the *Ensatina* salamanders living around the San Joaquin Valley in California? These fascinating creatures originated north of the valley and spread south along the valley's inland side and coastal side over millions of years before reuniting south of the valley. As the salamanders journeyed south on either side of the valley, they evolved into the various subgroups that can be found dotting the valley's perimeter. In most places, members of neighboring subgroups differ slightly but can interbreed. This seems to make them members of the same species. However, the two subgroups that coexist south of the valley are markedly different and incapable of interbreeding. This is because the two subgroups are descended from populations that made their way down opposite sides of the valley, gradually accumulating evolutionary differences that rendered them reproductively incompatible.[15] Are these 'end' subgroups therefore separate species? Or do they belong to the same species due to the existence of an interbreeding continuum? The conceptual boxes known as species are clearly ill-equipped to capture *Ensatina* salamanders.

The concept of species is hardwired into us from a young age, when we first learn that elephants have long noses while giraffes have long necks. Consequently, you might currently be scrambling to rationalize the blow dealt to this concept in the previous paragraph. Surely, you may be saying to yourself, the silly salamanders above are no more than a peculiar outlier in a sea of clearly defined species. But if you think about it, the concept of species is clearly a product of our imagination, since it is virtually useless when applied to organisms that are very different from us. Think, for instance, of non-sexually reproducing organisms. Sexual reproduction is a relatively modern evolutionary invention, having arisen a mere one or two billion years ago and failing to catch on among countless organisms. In fact, organisms as diverse as bacteria and certain roundworms reproduce asexually,[16] rendering the criterion of interbreeding wholly irrelevant. How, then, can we assign them species? And if species are no more than an artificial concept we use to describe ourselves and the organisms that resemble us, how can Darwinian evolution be summarized as 'survival of the fittest species'?

Back to square one. What are the elusive entities competing for survival? As we've seen, evolution is not about the survival of the fittest individual, group or species. But what does that leave us? In his landmark 1976 book *The Selfish Gene*, Oxford professor Richard Dawkins proposed an intriguing answer. Natural selection, he argued, operates directly on genes. In other words, evolution centers on the survival of the fittest *gene*. At first blush, this seems ludicrous. Genes are, after all, just short segments of DNA tucked away in cells, seemingly sheltered from the cutthroat competition of the natural world. But once genes are incorporated into Darwin's theory of evolution, everything falls into place. As we've seen, hereditary traits are determined by proteins, each of which is encoded by a gene. A gene encoding an advantageous trait, such as a larger beak on an island with large nuts, increases an individual's chances of surviving to adulthood. As an adult, this fortunate individual will replicate the advantageous gene and pass it down to many offspring, while individuals who have underperforming genes will struggle to reproduce and consequently pass down fewer copies of them. In the long run, natural selection cares little for individuals; it is solely concerned with preserving the best genes within them. Like shadowy puppeteers, Dawkins' genes are locked in clandestine competition behind the façade of finches vying for nuts.

* * *

Genes aren't just running the show – they invented it! The gene-centered view of evolution hypothesizes that life originated billions of years ago when a simple gene arose in the 'primordial soup', a term used to describe the aqueous mixture of organic substances thought to have been present in the ancient Earth's water bodies. This first gene was very different from the modern genes that encode our traits. In fact, it may well have been composed of something other than DNA. So what made it a gene? The answer is that it was a molecule capable of replicating itself. Though this may seem counterintuitive, the emergence of a simple, self-replicating molecule is not at all unlikely. Consider, for example, how the crystals of many minerals 'grow' by locking atoms into the same pattern over and over again. From table salt to diamonds, the natural world is brimming with self-replicating molecules.

Soon after the appearance of the first gene, its copies, or offspring, multiplied exponentially. Most, however, were not identical to the original gene, because no copying mechanism is completely error-free. The replication errors yielded a diverse population of genes, which were soon forced to compete for a dwindling supply of the atoms they needed to create copies of themselves. Some replication errors impaired a gene's ability to replicate, while others improved it by, say, speeding up the replication process or enhancing a gene's ability to bind the atoms necessary for replication. Genes blessed with favorable errors replicated furiously, filling the primordial soup with their offspring at the expense of their competitors.

Many eons later, genes 'discovered' that surrounding themselves with rudimentary membranes provided protection and a stable replication environment, paving the way for cellular life. Multiple genes eventually came to inhabit each cell, cooperating to increase replication efficiency by developing intricate mechanisms. And cells later aggregated to form multicellular organisms with advanced gene replication apparatuses known as reproductive systems. Through it all, genes have been looking out for themselves – and themselves alone. They are completely indifferent to the bodies they pass through in their endless replication cycles. Indeed, it could be said that our bodies are no more than temporary replication machines to be used and discarded as our immortal genes skip to the next generation.

Incredibly, some of our genes are so selfish, they can't bring themselves to cooperate with their cellmates. A cellular existence, you see, requires all genes to be replicated just once every cell cycle, right before the cell divides into two. Most of our genes have made peace with this mandated reduction in replication events, but some genes, known as *retrotransposons*, simply refuse to accept it. Instead, they replicate themselves throughout the cell cycle and insert their copies into new locations in our DNA,[17] yielding a genome riddled with retrotransposons.

The selfish nature of our genes explains many biological phenomena, from the sacrifices parents routinely make for their offspring to the development of cancerous tumors. Cancer, a leading cause of death, is a disease characterized by the excessive, uncontrolled proliferation of cells in a body. Set aside the gene-centered view of evolution, and this phenomenon is truly puzzling – why would our own cells betray us? But the genes responsible for cancerous proliferation did not swear an oath of fealty to the body they inhabit. In fact, they may sometimes wish to 'rebel' against it.

Look at it this way – every gene in a cancerous cell is a direct descendant of the first gene that arose in the primordial soup. For billions of years, the genes now found in a cancerous cell replicated themselves over and over again. When multicellular organisms first appeared on the scene, they developed a new class of cells. These cells, known as *somatic* cells, make up most of a multicellular organism's body, while the reproductive cells give rise to the next generation. In other words, somatic cells are a replicative dead end – their genes will only replicate themselves a limited number of times, and they will never make the leap to the next generation of organisms. The genes we see in a cancerous cell lucked out; they ended up in a chrono-logical series of reproductive cells, where they were free to keep replicating and skipping from generation to generation. This replication frenzy continued until one day they found themselves in a somatic cell. Several replications later, the cell entered senescence – a state in which all replication ceases. And so, our avid replicators were forced to stop replicating for the first time since the emergence of the first gene.

Genes do not have feelings, but I like to imagine these wronged genes simmering with resentment for decades as oppressive cellular mechanisms prevented them from replicating. So when an error known as a mutation accidentally freed them from their molecular chains, they (very understandably) leapt at the chance to resume what they had been doing since time immemorial. Thus they began to replicate with impunity, ignoring the signals designed to stop them and producing a fast-growing population of aberrant cells known as a tumor. The result, an often deadly disease, is the price we pay for trying to hold back these master replicators.

As we have seen, genes are not concerned with the health of a body they happen to inhabit. They seek only to propagate themselves, whether this can be achieved by reproduction or cancer. Philosophers have long debated the meaning of life, but when we contemplate the ephemeral nature of our bodies and the tyrannical control genes exert over them, it becomes clear that life itself was dreamed up by our genial genes.

CHAPTER 2

Pernicious Parasites

As we've seen, an organism's ultimate goal is the propagation of its genes. The only remaining question, therefore, is how it can best achieve this objective. The finches Darwin observed on the Galapagos Islands worked hard to pass down their genes, foraging for nuts and diligently building nests. But Darwin may have also unwittingly encountered an organism with a radically different gene propagation strategy.

In his diary, Darwin described being bitten by a 'great black bug' in Argentina. He later developed numerous health issues, including fatigue and chronic digestive symptoms, which were to plague him until his death in 1882. His mysterious illness has been the subject of much speculation, with everything from lactose intolerance to religious guilt being proposed as the cause of his suffering. But in 1959, Israeli parasitologist Saul Adler suggested that the 'great black bug', or kissing bug, that had bitten young Darwin infected him with *Trypanosoma cruzi*, the parasite responsible for Chagas disease.[1] To understand Darwin's illness in light of this hypothesis, we must first understand what parasites are, and what makes them so fundamentally different from organisms such as finches.*

Some organisms work hard to survive and reproduce, spinning silk, rolling balls of dung or building dams by the sweat of their brow. Parasites, on the other hand, survive by

---

* Not all of Darwin's finches are fundamentally different from parasites. The vampire ground finch (*Geospiza septentrionalis*) sometimes parasitizes other birds by feeding on their blood.

taking advantage of these hard workers. This behavior, collo-quially known as 'leeching off', is named after leeches, blood-sucking parasites that feed on the hard-earned nutrients in our blood. *T. cruzi*, the suspected bane of Darwin's existence, is an endoparasite, which means that it takes advantage of other organisms from within. It reproduces inside a kissing bug and is eventually excreted with its feces when the kissing bug bites a human. *T. cruzi* then enters the human host through the bite wound and invades its cells, siphoning the nutrients necessary to go on a reproduction spree. Left untreated, *T. cruzi* can cause long-term organ damage, leading to the decades of chronic Chagas symptoms that Darwin likely experienced.[2]

Despite his poor health, Darwin lived to the respectable age of 73. At first glance, this seems counterintuitive – wouldn't parasites have shortened his life? But put yourself in the parasite's shoes. Securing a host is a lot like hitting a jackpot, given that a host provides virtually endless resources and a predator-free environment. Killing this golden goose is clearly not in the parasite's best interests. That's why parasites will typically do just enough damage to obtain whatever they can from their hosts without killing them – at least not before they're ready to move on to another host. Yes, you read that right; many parasites infect multiple hosts over the course of their life cycle, often undergoing drastic transformations as they skip from host to host. This astounding feat of serial exploitation requires outsmarting host defensive mechanisms, adapting to life within hosts, shamelessly manipulating hosts and, of course, reproducing at the hosts' expense.

* * *

Let us begin with the imperative to outsmart host defensive mechanisms. When a microorganism such as *Trypanosoma cruzi* invades a human body, the immune system detects it and sends giant cells known as *macrophages* to gobble it up à la Pac-Man. This process is called *phagocytosis*, which is why the bubble that encases the ingested microorganism is called a *phagosome*. To break down the contents of the phagosome, the macrophage fuses it with a *lysosome*, an organelle containing digestive enzymes, forming a *phagolysosome*. Normally, a microorganism unlucky enough to end up in a phagolysosome is quickly broken down to its molecules. But when *T. cruzi* finds itself in a phagolysosome, it disarms it and escapes unscathed into the rest of the macrophage, where it reproduces repeatedly.[3] Thus, the macrophage, originally tasked with eliminating *T. cruzi* infections, is tricked into providing a safe environment for this cunning parasite to replicate just out of the immune system's sight. So advantageous is this arrangement to *T. cruzi* that rather than wait for a macrophage to gobble it up, the parasite often actively invades macrophages in a process dubbed *induced phagocytosis*.[4]

The malaria-causing *Plasmodium falciparum* parasite is also adept at evading host defensive mechanisms. It starts out in the salivary gland of a female *Anopheles* mosquito and is injected into the human bloodstream via a mosquito bite. Following a stint in the liver, *P. falciparum* quickly invades red blood cells to escape the immune system's detection. Once inside, the parasite feasts on the red blood cells' contents and increases their permeability, drawing in additional nutrients it can guzzle.[5] This allows *P. falciparum* to multiply like crazy over 48 hours, before bursting out and quickly invading new red blood cells. In other words, this parasite outsmarts the immune system by darting from cell to cell in what can be described as a microscopic game of hide and seek.

Once a parasite has successfully outsmarted its host's defensive mechanisms, it must find a way to survive – and reproduce – within the host. In short, it must adapt to life inside the host's often quirky organs. Let us examine, for example, the small intestine. Despite its, well, *smallness*, the small intestine is responsible for a whopping 90% of nutrient absorption in humans. Why is this seemingly simple tubular organ so much better at absorbing nutrients than the rest of the digestive system? The answer lies in its compelling structure. Imagine ripping this page out and rolling it into a tube. If mushy food is pushed through this tube, nutrients can be absorbed by the tube's wall. But how much absorption can we achieve in this manner? Flatten the page on a table, and you have your answer – the area of the page is the area available for absorption. This area, however, is intuitively inadequate for the lion's share of nutrient absorption in the digestive system. To increase the tube's surface area, one could, theoretically, use a page ripped from a newspaper, thus increasing the tube's diameter. This approach would definitely lead to more absorption, but it is also inherently wasteful, since a larger diameter would allow more mushy food to pass through, and most of that food would not come into contact with the tube's wall.

In other words, if we had a *wider* small intestine, we would fail to absorb much of the nutrients in the food we ingest – a clearly undesirable outcome. So how else could we increase the tube's surface area? Evolution has provided us with a surprising trick. If you look closely at the small intestine's wall, you can see that it looks nothing like a smooth page. In fact, it's *bumpy*. The little bumps, or *villi*, that line the intestinal wall are there to increase its surface area and provide nutrients with countless more opportunities to be absorbed. But why stop there? If you examine a *villus* through a microscope, you'll see that it, too, is covered with multitudinous tiny bumps, or *microvilli*. Taken together, the small intestine's villi and microvilli increase the surface area available for absorption by several orders of magnitude.

Given the small intestine's highly absorbent structure, it is difficult to imagine a parasite obtaining enough nutrients to survive, let alone thrive, in this organ. And yet, that is precisely what roughly 6,000 species of tapeworms do. In fact, tapeworms are so skilled at stealing nutrients from the small intestine, they can use them to grow to a shocking length of 50 feet! To understand how tapeworms have adapted to life in our small intestine, we must first understand what tapeworms are, and more importantly, what they once were. Tapeworms are descended from non-parasitic, or *free-living*, flatworms. This means that they used to have all of the genes required to survive out in the 'real world', unsheltered by the bodies of their modern hosts. When the ancestors of tapeworms started down the path to parasitism, something fascinating happened – they experienced a colossal loss of genes that are virtually universal in the animal kingdom.[6] Thus, they lost genes responsible for everything from basic biochemical processes to organ formation, reducing tapeworms to simpler creatures with a growing dependency on their hosts. Meanwhile, new genes evolved to facilitate the ever-increasing exploitation of these hosts. That's how tapeworms ended up with the strong suckers and hooks they use to cling onto our small intestine, but no mouth or digestive tract.

There is, however, one vestige of their flatworm ancestors that tapeworms have made a point of preserving, and that is their *flatness*. Adult tapeworms, though very long, are also exceedingly flat, rendering their bodies excellent surfaces for large-scale nutrient absorption. This is good news for tapeworms, given the loss of their mouth. But how can they compete with their host's small intestine, which is lined with villi and microvilli for optimal absorption? Incredibly, tapeworms too are lined with countless little bumps, enabling these sly parasites to feast on our food. In other words, to achieve maximal host adaptation, tapeworm evolution came to mirror host evolution, gradually transforming tapeworm bodies into inside-out intestines.

Some parasites adapt to life in different parts of a host so completely that they *speciate*, or branch off into distinct species that specialize in occupying specific body niches. Take lice, for example. Lice pierce the skin of their hosts and suck blood while clinging to hairs with the claws at the end of their six legs. All lice parasitize mammals or birds, with some poor animals hosting up to 15 different species of lice. Humans have long suffered from lice, as evidenced by the discovery of lice-infested mummies, but what of our earlier, and hairier, ancestors? As it turns out, the last common ancestor of humans and chimpanzees, who lived around 6 million years ago, was plagued by a single species of lice that roamed freely from head to hairy toes. But when humans began to lose their body hair, lice found themselves stranded on either the head or the pubic area. The two populations, separated by an unbridgeable hairless distance, gradually evolved into distinct species. Consequently, if a person is unfortunate enough to suffer from both head lice *and* pubic lice, the two kinds of lice cannot interbreed.

Incidentally, when humans began to wear clothes roughly 190,000 years ago, head lice gave rise to a new species of body lice. Despite their misleading name, body lice have very little contact with our (hairless) bodies. In fact, they live on unwashed clothes and often cling onto these clothes while sucking our blood. The distinctiveness of the three species of lice can therefore be said to rely on an inability to interbreed as well as behavioral differences. But what about physical differences? Did the bodies of the different species evolve to adapt to their host niches? The answer is a resounding yes. Examining the three species through a microscope reveals, for instance, that pubic lice have significantly larger claws than head and body lice. Pubic lice claws are so large, in fact, that these lice are colloquially known as *crabs*. But why did pubic lice evolve such huge crab claws in the first place? This is clearly an adaptation that enables pubic lice to swing, Tarzan-like, on pubic hair, which is much coarser than head hair.

* * *

Having adapted to their hosts, parasites are free to get down to business. This means exploiting hosts and manipulating them in any way necessary for maximal exploitation. Some parasites manipulate their hosts by stealing their genes and producing host-like proteins that commandeer the hosts' central nervous systems.[7] Others render their hosts unrecognizable. But why would parasites need to manipulate their hosts in the first place?

As we've seen, many parasites must pass through multiple hosts, belonging to different species, to complete their life cycle. This is known as a *complex life cycle*, and it can be surprisingly difficult to pull off due to the complexity of ecosystems. Consequently, many parasites resort to manipulating a host's behavior to increase their chances of being eaten by their next host. *Toxoplasma gondii*, for example, is an elegant parasite that invades host cells by building a molecular door in the cellular membrane, entering, and then taking care to shut the door behind it.[8] *T. gondii* often infects rodents such as rats, but can only reproduce sexually in cats. Having its rat host devoured by a cat is therefore this parasite's fondest wish. But unfortunately for *T. gondii*, rats will typically take great pains to avoid cats, (understandably) fleeing at the slightest whiff of cat urine. To solve this problem, *T. gondii* hijacks its host's brain and overrides its primal instincts, resulting in a rat whose curiosity is instead piqued by the smell of cat urine. But why, exactly, are infected rats so eager to explore areas tainted by this pungent odor? In an awe-inspiring feat of manipulation, *T. gondii* rewires the part of the rat brain that responds to the presence of potential sexual partners. Thus, when an infected rat smells cat urine, its arousal mechanism is switched on, tricking the poor rodent into feeling sexually attracted to its predator's urine![9]

Risky behavior can also be observed in other *T. gondii* hosts. Wolves, for example, either spend their whole lives with their birth pack, or strike out on their own to search for a mate. But why do some wolves choose to leave the safety of their families in favor of an uncertain future? As it turns out, when *T. gondii* ends up in a wolf host, it hijacks its brain and turns it into a daredevil that's 11 times more likely to leave its pack.[10] Were it not for this risky behavior induced by host manipulation, far fewer new packs would emerge, possibly threatening the species' survival. Incidentally, humans also become *T. gondii* hosts when they accidentally ingest oocysts, the parasite's extremely resistant eggs. This can happen, for instance, when a person cleans an infected cat's litter box and fails to wash properly before eating. The striking ease with which humans can pick up this parasitic infection explains why about 30% of us are walking around with *T. gondii* in our brains.[11] This, of course, means that if *T. gondii* can induce risky behavior in humans, we should see this behavior in just under a third of the population. But what kind of risky behavior do infected humans exhibit? Road rage, for example, is twice as likely to characterize infected drivers as compared to uninfected ones.[12] And intriguingly, *T. gondii* infection has been linked to a significantly increased tendency to found startups, one of the modern world's riskiest endeavors.[13]

Host manipulation can also serve purposes beyond routine host-hopping. A good example is the manipulation orchestrated by *Heterorhabditis bacteriophora*, a nematode worm that burrows into a caterpillar's mouth or anus, or directly through its skin. Once inside, the parasite kills its host and feasts on its tissues, founding a multi-generational dynasty in the caterpillar's corpse. Nothing, it seems, can stop *H. bacteriophora* from taking full advantage of its host – nothing save for predation, that is. You see, caterpillars are routinely gobbled up by birds, and if an infected caterpillar were to become some hungry bird's snack, the parasites within it would meet an untimely death. To solve this problem, these crafty parasites make their (dead) host glow red[14] – an alarming color predators find unappealing. Put simply, the parasite's power to manipulate its host is so absolute that even death can do little to weaken it.

* * *

Host manipulation is ultimately geared towards a parasite's greatest goal – reproducing at its host's expense. And when it comes to exploiting host resources for reproduction, parasites pull out all the stops, reproducing both sexually and asexually with staggering prolificacy. Why not be fruitful and multiply, you might be thinking, if the host is picking up the tab? This, however, is only part of the story. In fact, while parasites have access to seemingly infinite resources for reproduction, they are also compelled to reproduce fervently or risk extinction.

To understand why, let us turn to one of the quirkiest animals on the planet, the Australian echidna. Echidnas are egg-laying mammals with long, electroreceptive snouts. This means that their snouts can sense electrical currents, a handy skill they employ to locate prey. Oh, and they regularly blow mucus bubbles out of their snouts to keep cool.[15] This, of course, can hardly be expected to provide sufficient relief from the scorching Australian heat, which is why echidnas also *sploot*, or belly-flop on cold surfaces. Every year, an adult female echidna will lay a single egg into her pouch. That's right, echidnas have pouches – just like their koala compatriots! When the baby echidna, or *puggle*, hatches, it spends seven weeks in its mother's pouch before growing sharp spines and being evicted by its long-suffering parent. In short, echidnas beget just one little spiky, snot-bubble-blowing puggle a year.

But what about *Eimeria echidnae*, a parasite found exclusively in the digestive systems of echidnas? Does it echo its host's reproductive approach? Not by a long shot. An echidna will invest a great deal in a single puggle, going so far as to carry it in a pouch long after birth, to give the puggle a solid chance of reaching adulthood and passing on its genes. Parasite reproduction is a lot riskier. Will a given parasite's genes ever find their way to a new host, or will they disappear into oblivion when the current host dies? Pampering individual offspring will do little to guarantee their success. To ensure that at least one copy of its genes ends up in a new host, a parasite must constantly generate countless descendants. As a consequence, apparently healthy echidnas will shed up to tens of thousands of *E. echidnae* oocysts in a single gram of feces.[16]

Not all parasites live in or on their hosts' bodies. Some are more like that annoying friend who shows up empty-handed and ravenous to a potluck. About 40% of cuckoo bird species, for example, never build nests. Instead, they outsource their parenting by laying an egg in the nest of another bird and fooling it into feeding the cuckoo chick as though it were the host's own. To add insult to injury, the cuckoo egg hatches before the host's eggs, giving the cuckoo chick time to push the host's eggs out of the nest. Thus, the young parasite ensures that it will not have to share any of the food the host brings to the nest. Other parasites exploit hosts on their home turf. A good example is the 50 or so species of ants that have long lost the ability to run their own nests. To survive, these *slave-raiding ants* attack the nests of other species of ants and kidnap pupae and larvae. Back in the parasites' nest, the unsuspecting hosts hatch into slavery and spend their lives cleaning, caring for young, and feeding their masters mouth-to-mouth. So complete is their subordination that although they make up roughly 90% of the ants in a slave-raiding ant nest, they never attempt to overthrow their wily masters. Meanwhile, other parasitic ants take the opposite tack to host exploitation. Instead of carrying hosts back to their nest, *workerless social parasites* will slip into the nests of their hosts and just stay there, living out their lives in leisure.

It is all too easy to fall into the trap of viewing parasites as fascinating yet anomalous organisms. But parasites are far from being an anomaly in the natural world. Parasites were plaguing dinosaurs 100 million years ago.[17] Nowadays, parasitic diseases cause over a million human deaths every year,[18] and tens of billions of dollars are spent annually to combat livestock parasites. How do parasites manage to do so much damage? It's simple: they just make up the majority of living species, rendering parasitism the most popular lifestyle on Earth. Indeed, while humans are plagued by over a hundred species of parasites, we are far from alone – virtually all free-living species suffer from them. Parasites, too, can suffer from their own parasites known as *hyperparasites*. Put this way, it is evident that we free-living organisms are mere outliers in a world of pernicious parasites.

CHAPTER 3

Vivacious Viruses

Darwin was interested in grand, epic mechanisms. When he published *On the Origin of Species* in 1859, he unleashed a meticulously fact-based explanation of how huge changes occur in organisms over colossal scales of time. That same year, Louis Pasteur, a French chemist who focused on 'the infinitely small in nature', proved that the world is teeming with minuscule life forms. These *microorganisms*, he believed, were responsible for disease. Despite being invisible to the naked eye, microorganisms could be observed by Pasteur and his contemporaries through microscopes, paving the way for the methodological development of several vaccines. Pasteur himself, in fact, developed the first vaccine against rabies! But when he used his microscope to examine samples taken from rabid individuals, he failed to detect the microorganism responsible for the disease. Fortunately, Pasteur's associate, Charles Chamberland, developed a filter that removed bacteria from water in 1884. Pasteur used this filter in his experiments on rabies and found that bodily fluids passing through the filter retained their infectiousness. This, he reasoned, could only mean that rabies was caused by entities significantly smaller than bacteria, which could not be viewed through microscopes nor readily filtered out of liquids. But what exactly were these entities? Pasteur, the titan of microbiology, never found out.

Three years after Pasteur's death in 1895, a Dutch microbiologist named Martinus Beijerinck dubbed these tiny entities *viruses*, a name derived from the Latin word for poison. Beijerinck believed viruses to be liquid in nature, a hypothesis later disproven when the American biochemist Wendell Meredith Stanley demonstrated that viruses are particles. We now know that these viral particles cause diseases ranging from chlamydia to COVID-19, and that there are around 10^{31} viruses on Earth.[1] Yes, that's a one with 32 zeros after it – a number greater than the number of stars in the universe! And here's the kicker: viruses aren't technically alive. To understand why viruses have been described as existing 'at the edge of life',[2] we must first strive to understand what life is with a simple thought experiment.

Imagine, if you will, setting out to search the galaxy for alien life forms. Given that life may well look very different on other planets, how, exactly, will you know it when you see it? One sensible approach would be to come up with a list of characteristics shared by the life forms we know, and to use it as an astrobiological checklist. The problem is that, due to some delightful irregularities in the natural world, drafting such a list is much harder than it seems. Movement, for instance, appears to be an obvious characteristic of living things. But plants don't move, and they are undoubtedly alive! Moreover, some decidedly non-living things, such as the wind, are deceptively motile. Perhaps, then, it is the acts of feeding and growing that set us apart from inanimate objects. This sounds plausible, until one considers the several species of butterflies that are physically incapable of feeding, or the *Pseudis paradoxa* frog, which starts out as a giant tadpole and shrinks as it ages. We can also attempt to define living beings as things that reproduce, but then again, the vast majority of ants and bees lack a reproductive system (only the queen and a handful of males produce offspring). And as we've seen, crystals can be said to reproduce, rendering this criterion meaningless.

If all this is confusing, rest assured that life on Earth can also be defined by a much simpler condition – the existence of cells. Cells, the basic units of living beings, are microcosms of 'ideal' life forms. They move and possess myriad moving components. They feed and grow. And crucially, they reproduce by replicating their DNA and splitting into new cells. How do cells manage all of this? With the help of proteins, of course! Recall that proteins do everything from making up subcellular structures to catalyzing chemical reactions. Their pervasiveness would be impossible to achieve if cells weren't, at their core, protein factories.

Indeed, one could describe cells as zealously devoted to the process of stringing together amino acids in the order dictated by DNA. But how do cells 'read' a stretch of DNA codons in the first place? To answer this question, we must first familiarize ourselves with a molecule known as ribonucleic acid, or RNA. Like DNA, RNA is a chain composed of four types of nucleotides, enabling the storage of information in its sequence. But unlike DNA, RNA is typically *single-stranded* in cells. This means that RNA molecules are usually separate chains of nucleotides, whereas DNA molecules are *double-stranded*, or composed of two nucleotide chains linked all along their length. As we've seen in Chapter 1, each DNA codon encodes a certain amino acid in a protein's sequence. However, genetic information does not pass directly from DNA to proteins. To synthesize proteins, cells *transcribe* DNA sequences into RNA sequences, and then *translate* RNA sequences into protein sequences. Thus, RNA serves as a middleman in the production of the roughly 42 million proteins found in a single cell.[3]

Let us now return to viruses. How do we know that viruses aren't alive? The answer is that viruses cannot move, feed, grow or reproduce independently. Moreover, they aren't composed of cells. In fact, a typical virus is just a protein envelope surrounding a bit of DNA or RNA. How, then, do viruses proliferate, often causing devastating disease as they spread? It's simple: they infect cells and hijack their protein-manufacturing machinery – the very machinery that makes cells alive! If a virus happens to come into contact with the right kind of cell, it will inject its genetic material (DNA or RNA) into the cell and fool it into accepting the viral genetic material as cellular genetic material. The poor host cell is thus duped into replicating the viral genome and synthesizing the viral envelope proteins encoded within it. It then dutifully packs the new copies of the viral genome into new viral envelopes in the many *viroplasms*, or 'virus factories', that spring up inside an infected cell. In return, the young viruses proceed to kill the host cell as they simultaneously burst out of it to infect new cells. It should be emphasized that this astounding feat of perfect parasitism, in which absolutely everything required for reproduction is stolen from the host, is accomplished by a mere chain of nucleotides in a protein container. Indeed, as the esteemed philosopher Daniel Dennett once pointed out, a virus is a string of nucleic acid with attitude.

* * *

So, viruses are simultaneously simple *and* stupefyingly complex. Take, for example, *multipartite* viruses, which make up 35–40% of viral genera. The genome of these peculiar viruses is segmented, meaning that genes encoding different envelope proteins are carried by separate viral particles. It was once thought that all genome segments must converge in one cell for viral reproduction to take place, but the likelihood of this fortuitous occurrence is inversely proportional to the number of genome segments. The Faba Bean Necrotic Stunt Virus (FBNSV), for instance, is a multipartite virus with no less than eight different types of viral particles, each carrying a distinct genome segment. The probability of all eight FBNSV particles happening to infect the same cell seems vanishingly small, especially when you consider that only certain pea and bean plant cells can be targeted in the first place. So how does FBNSV manage to pull off the assembly of new viral envelopes? As it turns out, different FBNSV particles rarely infect the same cell. Instead, they infect multiple cells and orchestrate the exchange of viral proteins between them.[4] Thus the genomically simplest of viruses effortlessly commandeer whole networks of cells to mass-produce new viruses.

Multipartite viruses can be thought of as viruses that tear up and disperse their genetic user manual. Interestingly, some viruses dispense with their user manual entirely. A good example is the Hepatitis Delta Virus, or HDV. The HDV genome is a circular chain of RNA, which, oddly enough, does *not* encode the HDV envelope proteins. This means that if HDV infects a cell, it cannot trick it into synthesizing new viral envelopes, simply because it lacks the genetic information necessary to do so. But then how do HDV particles exist in the first place? How did their protein envelopes come to be? The answer is brimming with poetic justice. The Hepatitis B Virus, or HBV, is a 'normal' virus that encodes its own envelope proteins. In other words, when HBV infects a cell, viroplasms quickly form and commence HBV envelope production. And if that same cell is also infected by HDV, new copies of the HDV genome are summarily packaged into *HBV* envelopes. HDV can therefore be said to be stealing the very proteins HBV tricked the host cell into synthesizing. Since it can only reproduce in the presence of HBV, it is known as HBV's *satellite virus.*

Viruses, as we've seen, are little more than genetic material in protein containers. Given their simplicity, you'd think they would be eager to hold onto their few components. However, the existence of multipartite viruses and satellite viruses demonstrates the ease with which some viruses misplace their genetic material. Worse still, other viruses can be downright careless with their protein containers! Take, for example, the Herpes Simplex Virus, which is famous for periodically causing unsightly cold sores. During infection, the virus's linear DNA is injected into a cell, where it quickly enters the nucleus and circularizes to form an *episome.* But then something surprising happens – the episome may forget to leave its host! This happens when the Herpes Simplex Virus enters a state of latency, lingering in a cell for years as a container-less episome. Occasionally it will 'awaken' and take advantage of cellular mechanisms to generate new viral particles, whereupon cold sores may break out. This is why one can never be cured of herpes; its episomes persist for life.

When the Herpes Simplex Virus takes on the form of an episome, it is, in a way, crashing on the cell's couch. Remarkably, other viruses just move in with their unwilling host. The infamous Human Immunodeficiency Virus (HIV), which causes AIDS, is a fine example of an unwelcome roommate. Unlike 'simpler' viruses, HIV's envelope surrounds genetic material as well as a handful of proteins required for successful infection. And one of these proteins, known as *integrase*, actually integrates the viral genome into the cellular genome! Integrase pulls this off by brazenly clipping the cell's DNA and inserting viral DNA between the clipped ends, blurring the lines between the virus and its doomed host. As outlandish as this seems, HIV is far from alone – many viruses regularly integrate their genome into cellular DNA, and have been doing so since time immemorial. Consequently, a solid 8% of the human genome consists of ancient viruses that haven't seen the inside of a protein envelope in eons.

It goes without saying that cells do not appreciate being callously exploited by villainous viruses. This is why they evolve defensive mechanisms to keep viruses at bay. Viruses, in turn, evolve mechanisms to overcome these defensive mechanisms, forcing cells to evolve additional defensive mechanisms, which viruses must now evolve to overcome, and so on. In other words, viruses and their hosts are locked in an epic evolutionary arms race that puts the Cold War to shame. To understand just how intricate this arms race can become, let us turn to the Turnip Crinkle Virus, which infects *Arabidopsis thaliana*, a small flowering plant.

The Turnip Crinkle Virus genome consists of a single strand of RNA that is injected into *A. thaliana* cells. Recall that when viruses infect cells, they take advantage of host resources to replicate the viral genome. So how is the Turnip Crinkle Virus genome replicated? It's simple – the strand of viral RNA serves as a blueprint upon which cellular RNA nucleotides are arranged to create new strands of viral RNA. As we've seen, RNA is normally single-stranded in cells. However, when viral RNA is being replicated in an *A. thaliana* cell, it is (briefly) double-stranded. This is significant because it solves a big problem for the *A. thaliana* cell. In order to 'fight' the Turnip Crinkle Virus, the cell must first be 'aware' that it is under attack. Fortunately for the cell, viral RNA replication necessitates the temporary creation of an anomalous double-stranded RNA molecule, which signals that the cell has been infected.

The cell responds to the presence of double-stranded RNA by launching a counter-attack designed to prevent viral RNA from being translated into proteins. *A. thaliana* has evolved a protein known as Dicer-like protein 4 (DCL4) which chops up the double-stranded RNA, effectively thwarting the virus's vicious plans. Undeterred, the virus has evolved an envelope protein known as P38 that suppresses DCL4 by an unknown mechanism, thereby protecting viral RNA. This renders DCL4 useless, since it can longer be used to prevent viral RNA translation. But instead of evolving a novel approach to fighting the virus, the cell has tweaked DCL4 just enough to stay ahead in the arms race. Its new version, known as Dicer-like protein 2 (DCL2), can also chop up double-stranded RNA, sidestepping the bothersome P38 protein.[5] Although this last step gives the cell the upper hand, the arms race is far from over. The virus will undoubtedly eventually evolve a weapon to counter DCL2, and the *A. thaliana*–Turnip Crinkle Virus arms race will rage on until the end of time.

The legendary arms race between viruses and host cells is far from the only battle that viruses are busy fighting. In fact, different kinds of viruses will often fight *each other* over their most precious resource – hosts. Let us return to the Turnip Crinkle Virus, the bane of *A. thaliana*'s existence. The Turnip Crinkle Virus, as it turns out, is hardly the only virus seeking to exploit *A. thaliana* cells. In fact, it must often compete with the Cucumber Mosaic Virus for the use of cellular replication machinery. When this occurs, the P38 protein, which the Turnip Crinkle Virus uses to suppress a cellular defensive mechanism, is somehow wielded to ward off the Cucumber

Mosaic Virus. Consequently, when the Turnip Crinkle Virus infects *A. thaliana*, it renders the plant resistant to Cucumber Mosaic Virus infection.[6]

* * *

As a living being, the reader probably finds viruses extremely weird. Indeed, the ability that simple, non-living entities possess to manipulate each other as well as highly complex organisms is so astonishingly alien, their very existence is practically begging for an explanation. So where do viruses come from? And what were the earliest viruses like? We don't really know, but virologists have come up with three compelling hypotheses.

According to the *degeneracy hypothesis*, viruses used to be small cells that parasitized larger cells. Over time, they were reduced to their current form by a process known as parasitic degeneracy, in which a parasite becomes so dependent on its host for basic functions that it gradually loses parts and mechanisms required for independent living. Tapeworms, as we've seen, used to have a digestive tract, but as they adapted to life inside a host that does all of the digesting for them, their digestive tract atrophied. Consequently, tapeworms now acquire nutrients by absorbing them directly from the host's gut. Viruses, according to the degeneracy hypothesis, lost their cells in a similar process upon discovering that hosts can be tricked into doing absolutely everything for them.

The degeneracy hypothesis may sound far-fetched at first, but a microorganism discovered in England in 1992 seems to support it. At first, this microorganism, which was large enough to be visible under a light microscope, appeared to be a bacterium. However, when researchers failed to analyze its genome using a bacteria-specific technique, the truth emerged – the microorganism was actually a giant virus, or *girus*.[7] Dozens of girus discoveries followed, unveiling multiple girus families. But how did these colossal viruses, some of which exceed the size of typical bacteria, go unnoticed by generations of virologists? Recall Pasteur's use of Chamberland's filter to shed light on the size of viruses. This classic experiment established the practice of isolating viruses with filters containing pores that are simply too small for giruses. Thus, giruses were consistently excluded from virological studies before they began. Moreover, any giruses observed under light microscopes were mistaken for bacteria, allowing them to hide in plain sight for over three centuries.

Giruses, as it turns out, are remarkable viruses. They are parasitized by other viruses known as *virophages*, which, unlike regular satellite viruses, impede girus replication.[8] Girus genomes can also contain up to thousands of genes, surpassing many bacterial genomes. And girus DNA is wrapped around special proteins called *histones*, a feature once thought to be a hallmark of nucleated cells.[9] But the most surprising thing about giruses is the exceptional variety of their genes. Viral genes usually encode two basic types of proteins – the proteins found in viral particles, and the proteins required to overcome cellular defenses. Giruses, on the other hand, contain an assortment of genes responsible for decidedly nonviral functions, such as nutrient intake. Some giruses may even be capable of energy production, a metabolic feat previously believed to be exclusively cellular![10] Naturally, this renders giruses a perfect 'missing link' between run-of-the-mill viruses and their hypothesized cellular ancestors.

According to the *escape hypothesis*, viruses originated from genetic material that somehow escaped the confines of cells. Can such escapees turn into infectious particles? At first glance, this beggars belief. However, a simple experiment performed on an equally simple virus sheds light on how viruses may have first evolved. The Cowpea Chlorotic Mottle Virus is a spherical plant virus composed of just 180 identical envelope proteins and four genes encoded in RNA. In 2016, American researchers discovered that under certain conditions, simply mixing purified Cowpea Chlorotic Mottle Virus proteins and RNA in water yields infectious particles.[11] In other words, the

Cowpea Chlorotic Mottle Virus possesses the extraordinary ability to spontaneously assemble itself from scratch! This, of course, lends credence to the hypothesis that the first viruses were runaway genes which stumbled upon discarded proteins.

Finally, according to the counterintuitive *virus-first hypothesis*, viruses are actually the ancestors of cells. Think of it this way: cells, like viruses, are genetic material containers, albeit considerably more complex ones. And similarly to viruses, everything cells do serves the ultimate goal of replication. But how could viruses replicate themselves before there were cells to hijack? This was a head-scratcher until the genome of *Pithovirus*, an ancient girus discovered within 30,000-year-old Siberian permafrost, was shown to encode DNA replication proteins, suggesting that viruses were once capable of independent replication.[12] Viruses may therefore be relics of a bygone era before the emergence of cellular life, which means that cells are little more than glorified viruses. Perhaps, then, seeking to explain the existence of viruses is inherently misguided, given that long before the ascendance of cells, the story of life was being written by vivacious viruses.

CHAPTER 4

Machiavellian Males

A lthough *On the Origin of Species* is Darwin's best-known book, *The Descent of Man, and Selection in Relation to Sex*, published 12 years later, was even more widely read in Darwin's day. In this insightful book, which was surprisingly well received,[1] Darwin detailed his theory of *sexual selection*, which explains why organisms sometimes evolve traits that do not contribute to adaptation – or even hinder it outright. Take, for example, the sometimes outrageously colorful guppy. In the wild, female guppies are a dull gray, whereas males sport a variety of brightly colored spots, splashes and stripes. This may make male guppies aesthetically pleasing pets, but it also endangers them out in the wild, where bright colors render them conspicuous targets for predators.[2] If being so colorful actually reduces a guppy's chances of survival, why did male guppies evolve this trait in the first place? Wouldn't it make more sense for them to have evolved to resemble the drab females?

According to Darwin, some traits cannot be explained by *natural* selection because they serve to promote *sexual* selection. In other words, if a trait improves an individual's chances of mating, natural selection may take a back seat. In the guppy's case, females have been shown to prefer more colorful males as mates,[3] meaning that despite the inherent danger of swimming around in a clown suit, this trait is actually conducive to gene propagation. Similarly, many male animals invest costly resources in building and supporting massive weapons such as antlers and tusks. These weapons rarely contribute to survival, but they *are* wielded when males compete with each other over access to females.[4] Consequently, males who 'waste' a great deal of resources on these seemingly non-essential weapons are significantly more likely to reproduce.

Darwin's theory of sexual selection took the nineteenth-century world by storm. It was discussed in articles and speeches, and parodied in songs and cartoons. Respected authors were inspired to advise their readers to seek good-looking partners. And feminists seized upon the idea that females are nature's all-powerful mate choosers to rail against traditional gender roles and the practice of dressing up to attract men.[5] But why should *either* sex be forced to attract the other in the first place? Why do we even have sexes? Think about it: the earliest cells reproduced *asexually* by replicating their genes and dividing. Some organisms still reproduce this way, but many have embraced *sexual reproduction*, with distinct sexes and a gene-mixing process that precedes every generation. Why do we bother with sexual reproduction? What could possibly justify all those broken hearts and meddling in-laws when we could just split in two and call it a day? The answer is that asexual reproduction is often inadequate due to its production of lineages of organisms with virtually identical traits. This is adequate in stable environments, but more dynamic environments require species to constantly acquire new traits to survive. And what better way to acquire new traits than to mix genes? Sexual reproduction may be costly, but it is also invaluable because mixing genes generates the genetic diversity necessary for adaptation to an ever-changing environment.

Sexual reproduction is not the only way to mix genes. Bdelloid rotifers, tiny asexual animals found in freshwater habitats ranging from lakes to the film of water on mosses, have evolved an alternative gene-mixing method. When these remarkable creatures are subjected to drought, their DNA shatters into small fragments that reassemble upon hydration.[6] This reassembly, however, is far from perfect, and consequently, genes are routinely shuffled around the genome. Thus, an individual bdelloid rotifer can rearrange its own genes to generate genetic diversity. But that's not all – although bdelloid rotifers don't technically practice sexual reproduction, they do pick up genes from other organisms in their vicinity in a process known as *horizontal gene transfer*. These genes, incidentally, can originate in organisms as diverse as bacteria, fungi and plants![7]

Horizontal gene transfer, as it turns out, is a very popular gene-mixing method. It allows genes to flow from plant to plant,[8] from animal to animal,[9] and from bacteria to bacteria, fungi, plants and animals.[10] This brings us back to the original question of why we need sexual reproduction. Can't we just stick to asexual reproduction and the occasional horizontal gene transfer? Not really, because adaptation isn't just about acquiring beneficial genes; it's also about getting rid of *harmful* ones. If an asexual organism happens to acquire a harmful gene via horizontal gene transfer, all of its offspring will inherit this gene, and no matter which beneficial genes they go on to acquire, they'll still be stuck with this bad apple. In sexual organisms, on the other hand, genes are shuffled before every generation, and consequently, some offspring dodge the bullet of inheriting harmful ones. Sexual reproduction, then, is a great way to wipe the genomic slate clean of undesirable genes.

* * *

Having established the usefulness of sexual reproduction, we now turn to its key players – the sexes. Human readers are probably familiar with the concept of males and females, but snail readers may not be, because some snails are *hermaphrodites*, or organisms that produce both sperm cells and egg cells. Now, snails are by no means unusual; many other invertebrates are hermaphrodites, as are the vast majority of flowering plants. Hermaphroditism, as you can see, plays a central role in the natural world, a fact not lost upon Darwin, who believed that all living things are the descendants of a primordial unicellular hermaphrodite.

So how, exactly, does hermaphroditism work? *Simultaneous hermaphrodites* such as snails possess both male and female sexual organs,[11] which allow them to mate with all other members of their species – and sometimes even with themselves! *Sequential hermaphrodites*, on the other hand, start out as one sex and can later make the irreversible switch to the opposite sex. This may sound odd, but the ability to switch one's sex can actually be very useful under certain conditions.[12] The clownfish living in sea anemones is an excellent example of this. An anemone is typically home to a harem consisting of one large female, one reproductive male, and several small non-reproductive males. But if something happens to the female and the males find themselves alone, the reproductive male turns into a female and takes over for the missing matriarch. Thus, this sequential hermaphrodite ensures the survival of its harem for generations to come.

Hermaphrodites are extraordinary for two reasons. The first is that while other sexual species must find a suitable partner before they can reap the benefits of gene mixing, hermaphrodites can mate with *any* partner, and sometimes with themselves. The second reason is that hermaphrodites are the only organisms that uphold the value of egalitarian gene mixing. By ensuring that individuals are both male and female, or sometimes male and sometimes female, they maintain a utopian system in which sexual reproduction benefits everyone equally. This, as we shall see, is far from the case in species with distinct sexes. But first, let us delve into how distinct sexes arise in the first place.

Some organisms possess environmental sex determination systems. In alligators, for instance, the temperature at which an egg is incubated determines the sex of its hatchling.[13] This, however, is somewhat unusual, because sex is usually genetically determined. How does that work? Some organisms, such as swordtail fish, possess a *polyfactorial* sex determination system. This means that an individual swordtail's sex is determined by multiple genes which are dispersed throughout the genome.[14] This is unusual, too, because animals usually have specialized sex chromosomes.[15] Genomes, you see, tend to consist of discrete *chromosomes*, or multi-gene DNA molecules. Most chromosomes contain genes encoding a variety of proteins, but a special class known as *sex chromosomes* contain sex-specific genes. Consequently, the inheritance of these unique chromosomes is key to sex determination. So what kinds of sex chromosomes are there?

Some plants and algae produce spores containing either a U chromosome or a V chromosome. In this *UV sex determination system*, as it is known, the U chromosome produces a female, while the V chromosome produces a male.[16] Most sex chromosomes, however, work in more complicated ways. Organisms such as birds, some reptiles and some plants, for example, possess a *ZW sex determination system*. In this system, males have two Z chromosomes (ZZ) in each of their cells, while females have one Z chromosome and one W chromosome (ZW) in each of theirs. But when individuals produce *gametes*, or reproductive cells, things get interesting. Gametes contain half of a genome's chromosomes, which is why each of a male's sperm cells contains just one Z chromosome. A female's egg cells, on the other hand, contain either a Z chromosome or a W chromosome, which means that an offspring's sex is determined by its mother. If an egg cell containing a Z chromosome happens to be fertilized, the resulting ZZ pair of sex chromosomes renders the offspring a male. However, if an egg cell with a W chromosome ends up fertilized, the ZW offspring will be a female.

Intriguingly, some moth species have a *Z0 sex determination system*. This system differs from the ZW system in that only one kind of sex chromosome – Z – exists. Males have two copies of the Z chromosome (ZZ), whereas females have just one. Consequently, a female's chromosomes are denoted as Z-zero, or *Z0*. Since fathers can only produce sperm cells with a Z chromosome, sex is determined by mothers. So if a fertilized egg cell contains a Z chromosome, the ZZ offspring will be male, but if it lacks a sex chromosome, the Z0 offspring will be female.

This system is mirrored in the *X0 sex determination system* of many insects and a handful of mammals. In *this* system, the only existing sex chromosome is the X chromosome, of which males have one copy (X0) and females have two (XX). You can probably see where I'm going with this. Mothers always produce egg cells with an X chromosome, so if the fertilizing sperm cell contains an X chromosome, the XX offspring will be female, but if it lacks a sex chromosome, the X0 offspring will be male.

This brings us to the *XY sex determination system*, which characterizes humans and most other mammals. Similarly to females in the X0 system, females in the XY system have two copies of the X chromosome (XX) and a statistically unexciting egg cell production process. Males, of course, have one X chromosome and one Y chromosome (XY), and can consequently contribute an X chromosome to yield an XX female offspring or a Y chromosome to yield an XY male one. The XY sex determination can also accommodate outliers, such as male platypuses, who possess a jaw-dropping *five* pairs of XY chromosomes and produce XXXXX sperm cells and YYYYY sperm cells.[17] And although the XY system is by no means superior to the other systems discussed above, it is closer to our human hearts, which is why we will focus on it as we explore the unique role sex chromosomes play in evolution.

* * *

Sexual reproduction, as we've seen, was 'invented' to optimize gene mixing, thus accelerating adaptation. But when sex chromosomes appeared on the evolutionary stage, everything changed – including natural selection itself! Recall the dominant and recessive colors of Mendel's peas. We now know that dominant and recessive *genes* are behind these colors, and that this phenomenon is inextricably intertwined with sexual reproduction. As we've seen, our gametes contain half the chromosomes present in the rest of our cells, and consequently, when fertilization occurs, the newly formed individual gets two copies of each gene – one from the egg cell, and one from the sperm cell. If the two copies of a given gene are identical, the resulting trait is a no-brainer. But what if the two copies of a given gene encode different versions of a protein, as in the case of Mendel's green and yellow peas? This may result in an intermediate trait, or in a dominant gene's overruling of a recessive one, as in the case of the yellow pea gene overruling the green pea gene. In other words, a recessive gene can only exert its influence on a cell if it exists in two copies. If it shares a cell with a dominant gene, there will be no sign of its existence.

Recessiveness may sound insignificant when explained in relation to pea color, but it is actually a formidable stumbling block to adaptation. Think of it this way: if a highly advantageous gene happens to be recessive, it will often be invisible to natural selection. Put simply, although sexual reproduction benefits adaptation, it also makes it much harder for some promising genes to prove themselves. There is, however, one special chromosome where recessiveness is moot. I am referring, of course, to our dear X chromosome. Females, as we've seen, have two X chromosomes, which means that a dominant gene on one X chromosome will overrule a recessive gene on the other X chromosome. But males have just one X chromosome, which is why recessive genes that end up on it receive a rare opportunity to shine. This is not just hypothetical – when researchers analyzed the DNA of chimpanzees, they discovered that advantageous genes had accumulated on the X chromosome,[18] suggesting that this chromosome is a hotspot for evolutionary change. Migrating to the X chromosome, it seems, is the selfish gene's equivalent of moving to Los Angeles to make it big.

You'd think the Y chromosome, too, would boast more than its fair share of advantageous genes, but surprisingly, nothing could be further from the truth. Even though they evolved from the same ancestral chromosome,[19] the human X and Y chromosomes differ greatly – the X chromosome contains around 1,500 genes,[20] whereas the Y chromosome contains roughly 50.[21] In other words, the human Y chromosome is shrinking, and has already lost approximately 97% of its genes! Fortunately, this concerning shrinkage seems to have stabilized.[22] But if our Y chromosome does continue to shrink, it may well disappear entirely within a few million years, reducing the human XY sex determination system to an X0 sex determination system.

Incredibly, this wouldn't be the first time such a reduction occurred; research has ascertained that some species of mole voles[23] and spiny rats[24] lost their Y chromosomes and lived to tell the tale. If you find this hard to believe, you should know that sex determination systems aren't nearly as rigid as they appear. Multiple species of fish, amphibians and perhaps reptiles have undergone XY-to-ZW or ZW-to-XY transitions.[25] Moreover, species sometimes transition from environmental sex determination to genetic sex determination, from genetic sex determination to environmental sex determination, from hermaphroditism to having separate sexes, and from having separate sexes to hermaphroditism.[26]

* * *

If males are defined by V, ZZ, XY or X0 chromosomes, females are defined by U, ZW, Z0 or XX chromosomes, and the chromosomal composition of the two sexes is subject to evolutionary change, is it even possible to unambiguously define maleness and femaleness? Or are the terms 'male' and 'female' just random names for sexually compatible individuals? The answer is that although the mechanisms underlying maleness and femaleness differ among species, the core characteristics of males and females are universal. But to understand what makes a specific individual unequivocally male or female, we must return to the origins of sexual reproduction.

As we've seen, sexual reproduction started out as an easy way to mix genes. At first, both partners benefited equally from this arrangement, because both partners invested the same amount of resources in passing on the same number of genes. But as the two sexes took shape, sexual reproduction became increasingly *less* egalitarian. Let's start with the gametes produced by both sexes. Animal egg cells are usually much larger than the sperm cells that fertilize them. In humans, for example, egg cells aren't just larger than all other types of cells; they're also an astonishing 10,000 times larger than sperm cells![27] The reason for this disparity? An embryo will only implant in its mother's uterus about a week after fertilization, and the egg cell needs to store plenty of nutrients to get the growing embryo through this early period. The sperm cell, on the other hand, contains DNA, and very little else. This means that the production of each egg cell entails the investment of a great deal of resources, whereas sperm cells are ridiculously cheap to churn out. Consequently, females carefully allocate nutrients to just one egg cell every month during their childbearing years, while males think nothing of producing millions of sperm cells a day well into their golden years. In short, although the genes of males and females benefit equally from sexual reproduction, females pay a much steeper price to get the process going.

The first week after fertilization, of course, is just the beginning of the costliest of female endeavors – pregnancy. Creating a whole baby from a single cell, unsurprisingly, requires copious amounts of nutrients and energy. In fact, it is estimated that women invest a whopping 80,000 calories in each full-term pregnancy,[28] while men invest, well, nothing. But that's not all; pregnancy also takes another, more sinister toll on women's bodies. Research shows that pregnancies accelerate the process of cellular aging,[29] and can add as many as 11 years to a woman's biological age![30] Needless to say, men who father children do not suffer from comparable repercussions.

In an ideal world, males would pull their weight in sexual reproduction, or at the very least, feign gratitude for the enormous sacrifice made by females. But males seem incapable of this. Instead of gratefully accepting the colossal investment of maternal resources in their offspring, they actively push for mothers to invest *more*. We know this thanks to experiments conducted on the *Igf2 gene*, which plays a key role in fetal growth. When researchers disrupted the Igf2 gene in female mice, the female mice went on to bear normal offspring. But when researchers disrupted the Igf2 gene in male mice, the male mice fathered growth-deficient embryos.[31] To make sense of these findings, it is important to see pregnancy for what it really is – a struggle between a mother and her offspring. A mother, as we've seen, invests copious resources in her fetus, which translates into fewer resources for herself and for future fetuses. It is therefore in a mother's best interests to slow down fetal growth and conserve resources, which is why the Igf2 gene mice inherit from their mothers *does not* promote fetal growth. Consequently, the disruption of the maternal Igf2 gene goes unnoticed. Males, on the other hand, are rooting for fetuses in their struggle against their mothers. Since large fetuses are more likely to survive and propagate their genes, and since it is the females who foot the bill for this largeness, it is in the best interests of males to accelerate their offspring's fetal growth. That's why the Igf2 gene male mice pass down to their offspring promotes fetal growth, a phenomenon demonstrated by the stunted growth of embryos with a disrupted paternal Igf2 gene.

Having accelerated fetal growth at the expense of females, males turn to the next stage of reproductive exploitation – rearing young. This consists of various duties males typically shirk, such as lactation, which costs human mothers around 500 calories a day.[32] Females, it is clear, take on the lion's share of the resource-depleting activities required to raise healthy offspring, and even grand-offspring. Yes, you read that right – in many species, grandmothers provide young with invaluable support. The presence of a Killer Whale grandmother, for instance, increases a young Killer Whale's chances of survival, whereas a Killer Whale grandfather contributes nothing to his descendants' care.[33] Viewed this way, it is evident that we have strayed from the egalitarian roots of sexual reproduction, plunging headfirst into what can be described as reproductive parasitism. Females, by definition, are hosts, irreversibly parasitized from the moment of fertilization by Machiavellian males.

CHAPTER 5

Manipulative Mutations

According to Darwin's pangenesis hypothesis, hereditary traits arise as organisms adapt to their environments. Say, for example, that one of Darwin's finches found itself on an island with particularly tough nuts. Naturally, the finch was forced to strain its beak to crack the nuts it consumed, and over time, its beak grew stronger as a result. This altered the gemmules emitted by the finch's beak, and consequently, the finch's offspring inherited robust beaks. Darwin's hypothesis may sound plausible, but despite its undisputed elegance, pangenesis, as we've seen, is wrong. So how do different traits arise?

In Chapter 1 we encountered the first gene, a simple, self-replicating molecule. Most of the first gene's copies, or offspring, were not identical to it, because all copying mechanisms are prone to errors. If you find this hard to believe, grab a notebook and start copying this book into it by hand. Unless you work at an unreasonably slow pace, copying errors will start to crop up almost immediately. This may seem like a silly exercise, but before Gutenberg's 1440 invention of the printing press, all books were painstakingly copied this way – mostly by scribes and specially trained monks. Their inevitable mistakes yielded a variety of versions of ancient manuscripts, which historians often struggle to parse in their quest to uncover original texts and meanings.

Imagine, if you will, a monk bent over a manuscript in a dimly lit *scriptorium* for hours on end. What kind of copying error is he likely to make? He may, for instance, substitute a letter for a similarly shaped letter, especially if he is copying a text written in an unfamiliar language. This kind of copying error, or *mutation*, is actually very common in our DNA. Recall that DNA is composed of the nucleotides adenine (A), cytosine (C), guanine (G) and thymine (T). When our nucleotide sequences are copied in the course of genome replication, a handful of nucleotides are invariably replaced by other nucleotides. How can such single nucleotide substitutions affect us? To understand this, we must first gain a deeper understanding of how DNA encodes proteins.

As we've seen, in genes, each triplet of nucleotides, or *codon*, encodes a certain amino acid. Our four types of nucleotides can be arranged in 64 different codons, which is why you may be surprised to learn that our proteins are composed of a paltry 20 types of amino acids. So what are the 'extra' 44 codons doing there? Some of these codons serve as synonyms. This means that many amino acids can be encoded by up to six synonymous codons, just as many ideas can be expressed by synonymous words. When a single nucleotide substitution transforms a codon into a synonymous codon, we say that a *silent mutation* has occurred.

Not all codons encode amino acids. In fact, three of them are *stop codons*, or codons that signal the end of a protein-encoding sequence. Single nucleotide substitutions that transform amino acid coding codons into stop codons result in truncated proteins, and are consequently known as *nonsense mutations*. Surprisingly enough, not all nonsense mutations are disastrous. If a protein's functional domains are truncated, it cannot retain its functionality. But what if the new stop codon leaves most of the protein intact, shortening it by just a few amino acids? This kind of nonsense mutation may slip under the genetic radar.

Last but not least, some single nucleotide substitutions cause one amino acid to be swapped for another. These *missense mutations* may be inconsequential if an amino acid is replaced by another amino acid with similar characteristics. Alternatively, they can impair protein function by altering the characteristics of a key protein domain. And sometimes missense mutations *improve* a protein's function or drive it to acquire a novel function. Needless to say, novel functions may result in new traits that boost survival, and are therefore integral to adaptive evolution.

What other kind of copying error is our long-suffering monk liable to make? To speed things along, he probably doesn't copy one word at a time. Instead, he reads a sentence and jots it down quickly, before he can forget it. But after hours of committing passages to his short-term memory, the exhausted monk starts to slip up. Sometimes he accidentally inserts an extra letter, an extra word or multiple extra words into his copy. And sometimes he omits a letter or a word, or even accidentally skips a whole paragraph. In genetics, these mutations are known as *insertions* and *deletions*. Let us start with small insertions and deletions. What happens if a single nucleotide is inserted into or deleted from a gene's sequence during replication? What happens if a number of nucleotides are similarly inserted or deleted? Which is worse? The answer is that the insertion or deletion of a number of nucleotides is actually preferable, but only if the aforementioned number of nucleotides is *divisible by three*. To understand why, read the following sentence, which can be likened to a gene with five codons:

THE CAT ATE HER HAT

Assume a nucleotide in the middle of the gene is deleted. How will this affect the codons?

THE CAT AEH ERH AT

Oh, dear. What if a nucleotide is inserted, instead?

THE CAT ATT EHE RHA T

As you can see, single nucleotide insertions and deletions change all of the codons after the mutation. If these *frameshift mutations* occur at the beginning or in the middle of a gene, the encoded protein will be unrecognizable. But what if exactly three nucleotides are deleted?

THE CAT HER HAT

That's not so bad – in the absence of a frameshift, the codons after the mutation retained their meanings! If the missing codon encodes an unimportant amino acid in the protein's sequence, the mutation may even go unnoticed. And what if exactly three nucleotides are inserted into the gene?

THE BIG CAT ATE HER HAT

The insertion of a codon may very well enhance protein function. In some cases, it can even lead to the acquisition of novel functions, yielding new traits. The same, of course, is true for the deletion or insertion of multiple codons.

Copying ancient manuscripts all day, every day is grueling work. That is why our weary monk sometimes makes major copying errors, such as not noticing that two pages are stuck together and accidentally skipping a good length of text, losing his place and copying the same page twice, or even copying from the wrong manuscript. These large-scale copying errors occur in our DNA as well, and are known as *chromosomal mutations*. The DNA in our cells, as we've seen, is broken up into multiple discrete molecules known as chromosomes. Organisms differ in chromosome number; Jack Jumper Ants settle for just two chromosomes, humans are characterized by 46 chromosomes, and the microscopic *Sterkiella histriomuscorum* boasts thousands of tiny, single-gene chromosomes nicknamed *nanochromosomes.*[1] Before cells divide, they replicate all of their chromosomes simultaneously in a logistically complex operation that is practically begging for errors. That is why a long segment of DNA containing multiple genes can end up deleted from a chromosome, duplicated within the chromosome, or even flipped backwards. In some cases, a chunk of one chromosome is inserted into a second chromosome, and every so often two chromosomes will even swap chunks – just like the manuscripts of two disorganized monks sharing a desk.

If a gene encodes a protein with an important function, pivoting to a novel function may come at the expense of the original function, and may therefore prove disadvantageous, regardless of how promising the novel function may be. But when whole genes are duplicated in chromosomal mutations, the new copies of these genes are free to evolve novel functions, and are therefore key to evolutionary processes. Similarly, the deletion of genes may seem to be inherently catastrophic, but gene loss, too, can play a crucial role in adaptive evolution.[2] All mutations, then, from single nucleotide substitutions to massive chromosomal mutations, may be advantageous, disadvantageous or neutral.

Most mutations, however, don't get the chance to influence our evolution. This is because the vast majority of mutations occur in unimportant cells. If a beneficial mutation were to occur in, say, one of your bone cells, it would be meaningless in the grand scheme of things because it would never make it out of your body. For a mutation to have a real shot at evolutionary success, it would have to occur in a *gamete* – an egg cell or sperm cell. This way, it could be inherited by a child who would have a copy of that mutation in every cell, and who could pass it down to future generations. That is how evolution generally works; many small mutations occurring in gametes over many generations add up to new traits. But not all mutations are incremental. To elucidate how influential a mutation can be, let us explore single mutations that can change our lives – for better or for worse.

* * *

The first of these impactful mutations was key to humankind's harnessing of fire. If you are reading this book under an electric lamp, you are probably a member of an industrialized society which has minimized its reliance on fire in everyday life. Chances are, your only direct interaction with fire involves lighting (and blowing out) birthday candles. This is a pity, because our unfamiliarity with fire obscures the crucial role it played in our ancestors' lives. Fire wasn't just for staying warm in chilly caves; it was also extremely useful for scaring predators away, and it enabled our ancestors to remain active long after sundown.³ What did our ancestors do with this extra time? For starters, they took up cooking with fire. This proved particularly advantageous because cooking kills pathogens found in food, thus preventing all sorts of infections, and because cooking meat reduces the amount of energy necessary to chew and digest it, and consequently increases caloric intake by up to 78%.⁴ In addition to cooking, our ancestors started using fire to forge tools, which allowed them to exert ever-increasing control over their environment.

If fire sounds too good to be true, wait until you hear about its downsides. Aside from the obvious safety issues, burning wood produces smoke, which contains several toxic compounds. To stay warm, cook, forge tools or do anything else with fire, one must get close enough to inhale the toxic compounds. And once this unavoidable inhalation occurs, a protein known as the aryl hydrocarbon receptor, or AHR, binds to these compounds and shuttles them into the cell's nucleus, where their toxic effects can lead to death. At least, that's what happens in *other* vertebrates. The human AHR protein isn't very good at binding toxic compounds from smoke, and consequently, humans can inhale it relatively safely. So how did we end up with a fortuitously faulty AHR protein? This mystery can be solved by comparing the sequence of our AHR gene to that of our close (but extinct) cousins, the Neanderthals. The human AHR gene, as it turns out, contains a single missense mutation that weakens the protein's grip on toxic compounds, suggesting that after splitting from the Neanderthals roughly 800,000 years ago, our ancestors harnessed fire with the help of one swapped amino acid.⁵

All humans possess the mutated AHR gene, but many useful mutations can only be found in certain human populations. One such mutation allows a lucky minority to enjoy dairy delicacies and digest their star ingredient, lactose. Lactose is a *disaccharide*, or double sugar molecule, found in milk. Disaccharides can't be absorbed directly from the digestive tract; they must first be broken into *monosaccharides*, or single sugar molecules, by specialized digestive enzymes. Like all young mammals, human babies produce the *lactase* enzyme, which breaks down lactose in their mothers' milk. But while other young mammals stop producing lactase when it is rendered obsolete after weaning, some humans produce lactase throughout their lives. This *lactase persistence* enables adults to consume dairy products freely. *Lactose intolerant* adults, on the other hand, do not produce lactase, and if they consume dairy products, their inability to break down lactose results in symptoms such as abdominal pain, diarrhea and gas.

Lactase persistence is inextricably linked to geography. Thousands of years ago, when humans began experimenting with animal domestication, geographical constraints played an important role in the outcomes of these experiments. Ruminants such as cows, goats and sheep, for example, could only be raised on open grasslands, ruling out large swathes of land all around the world. Consequently, when humans first started milking ruminants, they only did so in certain geographical areas. And it was these humans who stood to gain from producing lactase in adulthood. So how did they manage to prolong lactase production? A clue lies in the DNA of Northern Europeans, who are largely lactase persistent. Sequencing the genome of Northern Europeans reveals that although their lactase gene is mutation-free, a single nucleotide substitution occurred just before the gene.[6] At first glance, this seems irrelevant – aren't we looking for a mutation *within* the lactase gene? But gene-flanking sequences are actually instrumental to the establishment of gene *expression* patterns, and their alteration can have far-reaching implications. It is therefore likely that this single mutation was responsible for the development of lactase persistence in Northern Europe, a process mirrored in other geographical regions, where other mutations have conferred the lactase persistence trait on unrelated populations.

As we've seen, mutations can help the inhabitants of certain geographical areas exploit local food sources. But mutations can do much more than that; they can also ward off *endemic diseases*, or diseases that consistently plague particular regions. Malaria, for instance, is one of the best-known endemic diseases. It is a constant threat in many tropical and subtropical countries, where it kills over 1,000 children every day. This devastation is caused by a parasite which invades red blood cells and consumes their contents. Now, due to malaria's endemic nature, the genes passed down in tropical and subtropical populations had countless generations to mutate in a malaria-stricken environment. If a protective mutation were to arise under these circumstances, it would, of course, be conducive to survival. But can a mutation do anything about a red-blood-cell-invading parasite?

Red blood cells are chubby little sacks stuffed with hundreds of millions of *hemoglobin* protein complexes, each of which is composed of two *α-globin* proteins, two *β-globin* proteins, and some iron. This is a good description of most people's red blood cells, but it does not apply to many sub-Saharan Africans, who inherit a normal β-globin gene from one parent and a β-globin gene with a single nucleotide substitution from the other parent. The single nucleotide substitution results in a missense mutation, which, in turn, results in a deformed β-globin protein that causes the red blood cell to collapse into a sickle-like shape. People of sub-Saharan descent who are *carriers*, or possessors of just one copy of this mutated gene, produce some normal red blood cells with the help of the normal β-globin gene, and some sickled cells. If a malaria parasite infects these carriers, it may invade a sickled cell and die, because unlike the oxygen-rich interior of normal red blood cells, sickled cells are characterized by dangerously low oxygen levels. Consequently, the sickle cell mutation protects nearly 90% of its carriers from severe or complicated malaria.[7]

It is easy to see why the sickle cell mutation spread quickly in sub-Saharan Africa. Eventually, however, the mutation's success turned a gift into a curse. This is because red blood cells don't just happen to be oxygen-rich; they are specialized oxygen couriers, filled with oxygen by our lungs and sent off to supply every part of our bodies with this vital resource. Sickle cell mutation carriers, as we've seen, produce both normal red blood cells and sickled blood cells, striking the sweet spot between meeting oxygen supply needs and providing malaria resistance. Naturally, as these carriers multiplied in the sub-Saharan population, many carriers ended up with carrier spouses. And unfortunately, the children of carrier–carrier couples have a 25% chance of receiving a copy of the sickle cell mutation from *both* parents, resulting in a condition known as *sickle cell anemia*. Sickle cell anemia is extremely dangerous because in the absence of a normal copy of the β-globin gene, *all* red blood cells are doomed to sicklehood, leading to chronic oxygen deprivation. In other words, possessing one copy of the sickle cell mutation is highly beneficial, but possessing two copies of the mutation is more lethal than malaria.[8]

The sickle cell mutation is such a perfect example of a double-edged sword, one would be forgiven for assuming that it is unparalleled in its ability to unleash both good and evil. But what about a hypothetical sex-determining mutation? Sexual reproduction, as we've seen, has resulted in both genetic diversity *and* grievous exploitation. But it hasn't always been around, which means that at some point in our evolutionary past, a mutation must have occurred that 'invented' males and females. What did this mutation look like?

The unassuming *Phycomyces blakesleeanus* fungus can shed light on this mystery. *P. blakesleeanus* fungi don't have fully fledged sexes, but they do have two rudimentary *mating types* which determine mating compatibility. Interestingly, the mating type of an individual fungus is determined by a single *locus*, or specific location on a chromosome. In some *P. blakesleeanus* fungi, this locus contains the *sex minus* (sexM) gene, and in others, it contains the *sex plus* (sexP) gene. That's it – that's the only difference between the two mating types! For humans, with their massive sex chromosomes, sexM and sexP genes are like snapshots of a distant evolutionary past. And the more we gaze at these intriguing snapshots, the more we wish we could turn the clock back even further, to the evolution of sexM and sexP in an asexual fungus. Fortunately, sequencing these two genes is a lot like turning the clock all the way back to their emergence. In fact, sequencing sexM and sexP reveals that they used to be the same gene! It seems that long ago, a mutation happened to flip this primitive gene backwards, accidentally creating an inverted version of it.[9] The original version and the inverted version of the gene then went on to accumulate mutations separately, gradually evolving into the cornerstones of sex.

* * *

Single mutations, it is clear, can achieve great things, such as enabling their bearers to harness fire, digest milk, ward off malaria and reproduce sexually. But there is a reason why the word 'mutation' is so often shrouded in negativity. Single mutations can, and do, wreak havoc on individuals, families and global populations. If the latter strikes you as an exaggeration, consider the case of a frameshift mutation that changed the course of history. This mutation seems to have arisen in the DNA of none other than Queen Victoria, who had nine children despite abhorring pregnancy and finding babies ugly. Once they were grown and presumably better-looking, Victoria arranged for her children and grandchildren to marry into various royal families, effectively uniting them into one big European royal clan. This might have been good for preventing wars (temporarily, at least) but it also resulted in a clotting disorder spreading throughout the royal bloodlines, because unbeknownst to Victoria, she was a carrier of a Hemophilia B mutation.

Hemophilia B, a disease characterized by an ever-present risk of excessive hemorrhaging, is caused by a mutation in a gene that encodes a clotting protein known as factor IX. Victoria probably had a frameshift mutation that resulted in the production of dysfunctional factor IX proteins, but was blissfully unaware of this problem because the gene that encodes factor IX is located on the X chromosome. This is significant because females have two X chromosomes in each of their cells, whereas males only have one. Since females have two copies of the factor IX gene, if one copy is mutated, the other copy can make up for it by producing functional factor IX proteins. Males, on the other hand, only have one copy of the factor IX gene, and if that copy happens to be mutated, they will suffer from Hemophilia B. Consequently, males are more commonly afflicted with Hemophilia B, while females like Queen Victoria are carriers of the disorder who can pass it on to their sons without personally suffering from the disease.

Victoria passed her factor IX gene mutation on to three of her children, including her daughter Alice, who in turn passed it on to her daughter Alexandra, who married Tsar Nicholas of Russia and had four daughters. While this might sound wonderful to modern ears, her apparent inability to produce a male heir was cause for great concern in the imperial court and contributed to her unpopularity. So when a son was finally born in 1904, the Tsar and Tsarina thought they were out of the woods – until little Alexei's umbilical cord was cut and his navel bled for hours. Terrified that the news of Alexei's hemophilia, which could render the most minor injury fatal, would endanger both Alexei and their family's future, the Tsar and Tsarina kept their son's disease a state secret. They became entangled with a faith healer and mystic who claimed he could cure Alexei, and made some disastrous political mistakes at his advice. Consequently, the Russian Revolution broke out, resulting in the murder of the imperial family and the establishment of the Soviet Union, the first communist state in the world. Put simply, a single frameshift mutation on a little boy's X chromosome brought about a revolution that changed the world forever.[10]

What are the odds of a single mutation sparking a revolution? Pretty high, as it turns out. In fact, some bees possess a revolutionary mutation that drives them to dethrone their version of the Tsar – the queen bee. Normally, worker bees leave all

That's a bit of a stretch!

procreation to the queen and dedicate themselves to lives of servitude. Although they are technically female, worker bees avoid reproducing and spend all their time cleaning, repairing and guarding the hive, foraging, producing honey, and taking care of the queen and her offspring. However, some South African Cape Honey Bees exhibit the bizarre *thelytoky syndrome*, which causes workers to disobey the queen and start reproducing independently. Amazingly, this act of open rebellion, which upends bee society, is caused by a single missense mutation.[11]

Mutations can also exert their devastating effects by altering the spatial structure of our genes. DNA molecules are usually double-stranded molecules, which means that every chain of nucleotides is linked to another chain all along its length. The two chains are twisted around each other, yielding the famous double helix depicted in everything from textbooks to Hollywood movies. But DNA doesn't always form a double helix, as the curious case of the *frataxin gene* demonstrates. This gene contains a series of consecutive guanine-adenine-adenine triplets known as *GAA trinucleotide repeats*. Most people have a handful of GAA trinucleotides in their frataxin gene, but a duplication mutation can result in more than a thousand of these repeats, and this expansion is at the root of a neurodegenerative disease called *Friedreich's ataxia*. What, exactly, causes Friedreich's ataxia? Is the frataxin protein truncated, or perhaps mutated beyond recognition? Neither, actually; the problem is that no frataxin protein is produced in the first place. The reason? Too many GAA trinucleotides cause the normally double-helical DNA to transform into *triple*-helical DNA, blocking access to the frataxin gene and preventing the cell from creating the protein encoded within it.[12]

The list of single mutations that cause terrible diseases goes on and on, but truth be told, they make up a minority of harmful mutations. They lead to immense suffering, of course, but in order to suffer, one must be alive, which means that these mutations cannot be lethal – or at least, not initially. But what about the potentially infinite number of lethal mutations that can occur? Given the sheer number of genes involved in embryonic development, it stands to reason that the vast majority of harmful mutations affect genes involved in various stages of this process, leading to miscarriage. This would explain why only about half of all egg cell fertilizations result in recognized pregnancies.[13] It would also explain why none of us are born with truly insurmountable copying errors, somewhat obscuring the fact that we are all at the mercy of our manipulative mutations.

CHAPTER 6

Crippling Constraints

Darwin did not believe that organisms could evolve in limitless ways. In *On the Origin of Species* he wrote about the law of 'Unity of Type', according to which organisms in the same class are always of the same *type*, regardless of their adaptive needs. Think, for example, of different organisms belonging to the class of mammals. Despite occupying a wide variety of ecological niches, their many adaptations never alter their basic type, which we can define with characteristics such as warm-bloodedness and lactation. In other words, evolution does not mean that the slate is wiped clean for every generation. Mutations act on organisms with pre-existing characteristics, which limit evolution at every turn.

This concept can be simultaneously obvious *and* difficult to grasp due to the ostensibly infinite power of mutations. Fortunately, it can be demonstrated with the help of a drowning whale. Yes, you read that right – whales can actually drown! Whales must surface to breathe, so when they are prevented from surfacing by, say, a fishing net or a giant squid, they drown. But why do they insist on surfacing at all? As aquatic animals, couldn't whales have evolved to breathe underwater by now? Underwater breathing mutations have certainly had enough time to occur, and their notable absence suggests that whale evolution is somehow limited.

All life was once aquatic. Hundreds of millions of years ago, animals began to inhabit shallower waters, and eventually, some of them ventured onto dry land. As they adapted to life on land, these newly terrestrial animals lost their gills and evolved lungs to extract oxygen from the air. Incredibly, some terrestrial animals *returned* to the water around 50 million years ago and evolved into aquatic mammals such as whales.[1] Who were these fickle whale ancestors? Darwin hypothesized that they were swimming bears who spent increasing amounts of time in the water, but we now know that whales are more closely related to hippopotamuses.[2]

Imagine, if you will, a primordial proto-hippopotamus wading into a lake. For some reason, it is in its best interest to stay in the lake for longer than its parents did. Perhaps a new predator has taken to terrorizing it on land, or perhaps more nutritious food can be found in the water. Either way, the primordial proto-hippopotamus is inclined to remain in the water, and any mutations allowing its descendants to adapt to this new lifestyle will be immensely advantageous. So what kind of mutations would we expect to see?

The primordial proto-hippopotamus had a mammalian respiratory system, which was gradually tweaked by mutations to facilitate aquatic life. Thus, the opening of the nasal passage moved to the top of early whales' heads, yielding a blowhole that allowed them to breathe while mostly submerged. Similarly, whales developed increased oxygen absorption capabilities that allowed them to spend longer periods of time underwater before needing to surface to breathe. These adjustments may strike you as odd – wouldn't reacquiring gills be more efficient? Well, yes, but that's just not how evolution works. Since no mutation can replace lungs with fully fledged gills in one fell swoop, the best mutations can do is improve the organs whales already have. The existence of lungs is, in other words, an evolution-ary constraint which mutations must work around. And that is how incremental changes to the mammalian respiratory system resulted in an aquatic animal that can drown.

* * *

Evolutionary constraints can also explain why we have yet to see a real-life Dumbo, the flying elephant. Elephant migration is a tedious business, especially during Africa's infamous dry season. Wouldn't it be easier if elephant ears evolved into wings, allowing elephants to fly around the savanna looking for food and water? Of course it would, but reasonably sized wings require much lighter bodies. Birds, for example, are significantly smaller, and therefore lighter, than elephants. Birds have also evolved hollow bones and hollow feather shafts to keep their weight to a minimum. Needless to say, hollow bones couldn't possibly support the massive frames of elephants. To solve this problem, Dumbo could, theoretically, evolve colossal ear-wings, but these wings would have to be so huge that their weight would render poor Dumbo immobile.

Similarly, evolutionary constraints are at the heart of our intellectual shortcomings. Have you ever wondered why humans, or any other organisms for that matter, aren't particularly smart? Super-level intelligence would surely give organisms an evolutionary advantage, so why hasn't it arisen? The answer lies in brain size constraints. Although brain size is not the only intelligence-determining factor, it is a good predictor of cognitive ability.[3] Consequently, instead of asking why no species is super intelligent and getting bogged down in arguments over the definition of intelligence, we can focus on a simpler question – why don't organisms evolve gigantic brains?

Sperm whales hold the record for the biggest brains in the animal kingdom. Their brains can weigh up to 20 pounds, but surprisingly enough, this does not make them geniuses. This is because unlike animals with small bodies, the sperm whales' leviathan bodies send their brains copious sensory signals for processing. Put simply, the bigger the body, the bigger the brain required to process the signals it generates. Consequently, it is not the brain's *absolute* size that determines an animal's intelligence, but rather its *relative* size. Humans, unsurprisingly, boast the largest brain-to-body weight ratio in the animal kingdom.[4]

Our brains have tripled in size over the course of our evolution from earlier primates, and are currently roughly six times larger than one would expect, given the size of our bodies.[5] This, however, is as big as they're going to get, for two simple reasons. The first is that our brain consumes roughly a fifth of our body's oxygen supply,[6] and we simply cannot afford to give it any more. We could, of course, evolve bigger lungs to increase our oxygen intake, but this would prove counterproductive because we'd be forced to evolve bigger bodies to house the bigger lungs, thus decreasing our brain-to-body weight ratio. Secondly, our oversized brains result in oversized heads, which make childbirth considerably harder and more dangerous than it is in other primates. Even larger brains would increase infant mortality, and would therefore be decidedly disadvantageous. So, if you've ever failed a test, rest assured that it's not your fault – it's the oxygen supply constraint and the childbirth constraint that are holding you back from intellectual greatness.

If we can't be geniuses, can we at least live forever? Humans have been seeking the fountain of youth for millennia, but unfortunately, death is another non-negotiable foisted upon us by an evolutionary constraint. The problem starts with the air we all breathe. We need oxygen to survive, but when we metabolize it, we produce by-products known as *reactive oxygen species*, or ROS. These troublesome molecules wreak havoc on our cells, irreversibly damaging everything from membranes to DNA. As the years go by, ROS-incurred damage accumulates, resulting in the decreasing functionality associated with aging.[7] This, of course, ultimately leads to the unavoidable death that awaits us all. In short, the most ironic constraint seems to be that we cannot evolve to be immortal because we must breathe.

Some evolutionary constraints are actually environmental constraints. Why can't hollow bones support an elephant? The real constraint in this case is the Earth's gravity, which dictates the minimum bone density required to support an elephant's weight. Similarly, the elements found in our environment limit the kinds of bodies we can build, and are therefore key evolutionary constraints. Alien elephants could, hypothetically speaking, end up with hollow bones made of ytterbium if they happened to evolve on a much smaller, ytterbium-rich planet. Keep *that* in mind next time you visit the zoo!

Environmental constraints can sometimes be boring. For instance, an environment that doesn't change too much constitutes a major evolutionary constraint, because adapting to such an environment requires very few changes. Put simply, stable environments may penalize organisms undergoing any significant evolutionary change, resulting in species that barely change at all. This, however, does not mean that extremely unstable environments are necessarily conducive to evolution. Environments that change *too* fast often lead to extinction, as in the case of organisms wiped out by invasive species,[8] or drastic changes in weather.[9] Evolutionary processes, it seems, thrive in the sweet spot between boring constraints and overly dynamic environments.

* * *

A particularly interesting class of constraints consists of evolutionary trade-offs, in which two traits limit each other's evolution. Take, for example, the replication speed trait. If our genes' only objective was to create copies of themselves, we would expect to see cells replicating their DNA feverishly. But DNA replication is actually a drawn-out affair, typically taking about eight hours in mammalian cells.[10] Why are our genes taking their time replicating? Because they aren't interested in merely creating copies of themselves; they're interested in creating *high-quality* copies that can carry on their replication mission. Replicating as fast as possible, you see, comes at a cost – mutations can accumulate quickly, rendering genes dysfunctional in just a few replication cycles. The only alternative is to slow down, proofread the replicated DNA, and correct mutations diligently, as specialized cellular systems do.[11] So accelerating replication in an attempt to dominate the playing field impairs accuracy, while maintaining accuracy necessitates slowing down the replication process. In other words, DNA replication is a trade-off between replication speed and replication accuracy.

Similarly, if reproduction is our endgame, we should expect to see organisms producing countless offspring. But if you look around, organisms such as humans seem to have a rather limited number of young. This is counterintuitive, suggesting that a trade-off is at the root of our reproductive strategy. But what could possibly be more evolutionarily advantageous than increasing the number of one's offspring, assuming that the offspring's DNA was copied accurately? High-quality DNA, as it turns out, isn't all it takes to succeed in life. Longer pregnancies, for example, yield highly developed young who are more likely to survive in the wild. Moreover, feeding and protecting one's young increases their chances of reaching adulthood and reproducing. Long pregnancies and caring for young, however, require the investment of parental resources that could otherwise be invested in producing more offspring. And therein lies the trade-off: one can either produce many offspring and invest so few resources in them that most are unlikely to survive – as insects do; or produce few offspring and invest heavily in them, ensuring that a significant proportion will pass on their genes – as humans do. Producing many offspring *and* giving them all a leg-up is simply not possible given the finite resources at our disposal.

Pathogens such as bacteria invest very little in their offspring. That is how they manage to multiply so quickly. But if pathogenic bacteria are so good at proliferating themselves, shouldn't we all be dead by now? What's holding them back? An evolutionary trade-off, of course! For pathogenic bacteria, proliferation isn't just a matter of reproduction; it's also a matter of contagious-ness, or the ease with which they can infect new individuals. And contagiousness is at odds with lethality, meaning that successful pathogenic bacteria cannot be both highly contagious *and* highly lethal. Think of it this way: highly lethal bacteria rarely get a chance to infect new individuals, because they kill off infected individuals before giving them an opportunity to interact with others. The most contagious bacteria, on the other hand, opt to keep their victims on their feet, mingling with others and spreading pathogens far and wide.

* * *

When discussing evolutionary trade-offs, it is all too easy to view the traits locked in these infinite battles as obligatory features of life. But these traits, like all traits, only exist because mutations happened to give rise to them. And if mutations cannot occur, new traits cannot develop, rendering the dearth of mutations a major evolutionary constraint. This constraint is far from hypothetical, because mutation rates vary widely between species. Viruses boast the highest mutation rates, which allow them to evolve new traits at lightning speed. Some species, on the other hand, barely mutate at all. In fact, Darwin himself coined the term *living fossils* to describe extant creatures that are virtually identical to their fossilized ancestors. The uneventful evolutionary history of living fossils does not necessarily mean that they are uniquely capable of copying their DNA very accurately. It could just mean that living fossils have evolved overly efficient mutation-correcting mechanisms, which are now inadvertently limiting evolution.[12]

Is evolution given free rein when cells do not crack down on mutations? Not always. To understand why this is the case, we turn to the quirky way in which bacteria acquire their genes. As we saw in Chapter 4, bacteria don't just inherit the genes of their ancestors, as we do; they also frequently swap genes with their neighbors via horizontal gene transfer. This is of great consequence because it means that in addition to evolving genes from scratch through mutations, bacteria also routinely gobble up ready-made genes from their environment, providing us with a unique opportunity to explore how genes that evolved separately behave when thrust together in a cell.

In one fascinating study, the genomes of thousands of strains of *E. coli* bacteria were analyzed to shed light on how new genes were acquired. The results revealed that the genes picked up via horizontal gene transfer were far from random – in fact, there appear to be hard rules governing their uptake. If, for instance, a certain gene, X, is present in a bacterium's genome, another gene, Y, is likely to be present in its genome as well, whereas a third gene, Z, is never found in the same genome as X. Clearly, some genes, such as X and Y, are compatible, while genes such as X and Z are incompatible. The acquisition of new genes is therefore dependent on the genes already present in a genome,[13] which means that a bacterium's existing genes limit its future evolution.

Last but not least, a gene's length may limit its evolution. Take, for example, the genes that encode proteins found in and around neurons. A great deal of neuronal genes, it turns out, are massive sequences spanning hundreds of thousands of nucleotides. Interestingly, many of these mega genes existed long before the emergence of neurons, and gradually morphed into neuronal genes as multiple lineages evolved nervous systems independently. This suggests that complex systems, such as nervous systems, require pre-existing giant genes to evolve.[14] And if an ancestral gene's length can prevent its future carriers from evolving complex features, then absolutely everything about us is a product of crippling constraints.

CHAPTER 7

Curious Characteristics

D arwin, as we've seen, was unfamiliar with DNA, though he did envision gene-like particles he dubbed *gemmules* being passed down from generation to generation. It probably never occurred to him that parents could pass down anything but gemmules to their offspring, because, well, why would they? The notion that some hereditary material does not determine any traits seems to defy logic. The truth, however, is that DNA is not just a sequence of genes. If it were, genome size would depend on gene number, but there doesn't seem to be any correlation between the two.[1] In fact, less than 2% of the human genome encodes proteins![2] So what is the vast majority of our DNA for?

Soon after it came to light that very little of our DNA is composed of genes, the term *junk DNA* was coined to describe the non-coding, and therefore presumably redundant, parts of our genome.[3] However, although most organisms carry around non-coding DNA, the perceived proportion of 'real' junk began to shrink when it emerged that non-coding DNA plays a variety of vital roles. Many non-coding sequences, it turned out, carry out important tasks such as increasing or decreasing gene expression, facilitating DNA replication or ensuring proper cell division.[4] Consequently, if junk DNA is defined as DNA that serves no purpose whatsoever, its known quantity decreases with every new discovery of non-coding DNA functionality, and may one day diminish to zero.

The existence of junk DNA in our cells, although bizarre, is hardly our only genetic quirk. Comparing the genomes of different species, for example, reveals that species differ not just in the genes they carry, but also in the *nucleotide composition* of their genes. As we've seen, DNA is composed of four nucleotides, colloquially known as A, C, G and T. At first glance, their prevalence seems unimportant. They are, after all, just letters in the DNA alphabet, and different letters can be used to construct synonymous words. But unlike the letters used to write this book, the letters used to write DNA sequences possess a deeper, almost mystical meaning. The nucleotide composition of DNA, as we'll see, is far from random; and more importantly, it exerts a profound influence on evolution.

Genes, it turns out, tend to have a higher *GC content* than the rest of the genome, which means that the nucleotide composition of genes is biased in favor of G and C nucleotides over A and T ones. GC-poor stretches of DNA, on the other hand, usually lack genes.[5] How did this peculiar state of affairs come to be? Research suggests that GC-poor genomes were gradually converted into GC-rich ones in an evolutionary process driven by what can be described as a biased molecular handyman.[6] DNA, you see, is prone to breaking, and when it does, cellular mechanisms quickly patch up gaps with nucleotides. But for some reason, many organisms, including humans, employ GC-biased repair mechanisms, which lead to the replacement of missing A and T nucleotides with G and C nucleotides. This replacement often comes at our expense,[7] because as we've seen in Chapter 5, swapping nucleotides can impair gene functionality. In other words, although natural selection favors advantageous mutations, repair mechanisms promote GC enrichment regardless of whether it results in advantageous or disadvantageous mutations.

Incredibly, once a genome's GC content is high enough, it can start influencing that genome's evolution. In one particularly eye-opening study on the subject, researchers constructed three versions of a certain yeast gene. All three versions encoded the same amino acid sequence, but by leveraging the silent mutations discussed in Chapter 5, the researchers were able to engineer a gene with low GC content, a gene with intermediate GC content, and a gene with high GC content. The three genes were then introduced into yeast cells and allowed to evolve freely. When their mutation rates were later analyzed, it emerged that the gene with high GC content was mutating much faster than the other versions,[8] indicating that GC content was pulling the strings behind evolutionary processes.

That's just the tip of the iceberg for GC content. Another revealing study uncovered a strong correlation between the number of genes in bacterial genomes and their GC content, suggesting that genomic expansion is associated with an increasing GC content.[9] Moreover, organisms characterized by large populations tend to boast genomes with high GC content,[10] implying that the GC content of DNA is linked to realms far transcending its molecular circumstances. Last but not least, GC content can reflect how organisms handle the reproductive trade-off discussed in Chapter 6. A remarkable study found that *unicellular*, or single-celled, organisms with high GC content invest very little in a large number of offspring, whereas unicellular organisms with low GC content invest a great deal in fewer offspring.[11]

GC content can also influence protein production. As we've seen in Chapter 5, many of the amino acids that make up proteins are encoded by different, *synonymous* codons. Given their synonymity, you'd expect these codons to appear at more or less equal frequencies in genes, but surprisingly, they rarely do. Think of it this way: nobody uses synonyms at equal frequencies! Most people employ biased lexicons based on their cultural, educational and professional backgrounds. Similarly, although genomes could theoretically use a variety of codons to encode the same amino acids, in practice they tend to favor certain ones. This phenomenon is called *codon usage*, and it can result in significant yet invisible differences between species. The lysine amino acid, for instance, is encoded by both the AAA codon and the AAG codon. Some species, like the probiotic *Lactobacillus acidophilus* bacterium, use the two codons interchangeably. Others, like the *Streptomyces venezuelae* soil bacterium, demonstrate a strong preference for the AAG codon, while a third group of species, which includes the *Buchnera aphidicola* aphid bacterium, exhibit an AAA codon bias. Why does this matter? Because codon usage influences the speed and accuracy of the protein production process. Differences in codon usage, in fact, can yield 15-fold differences in protein production efficiency.[12] And here's the kicker – the GC content of genomes determines their codon usage![13]

* * *

All of the aforementioned genetic quirks pertain to the nucleotides that make up DNA. But if you step back to see the big picture of DNA, nucleotides suddenly appear to be little more than an afterthought. To do DNA justice, we should really be focusing on its spatial structure, not its sequence. Take, for example, the fascinating subject of DNA ends. Some DNA molecules, such as the chromosomes of the human genome, are linear, which means that each DNA molecule has two ends. Consequently, each of these molecules must be replicated separately. This may sound insignificant, but the truth is that for species with many linear chromosomes, replicating all of them separately but simultaneously is extremely complicated.

Replicating linear DNA is also problematic because the ends of linear DNA molecules actually get a little shorter with every replication cycle.[14] This poses a significant issue since as we've seen, gnawing at the end of a gene can have deleterious effects. Put simply, our genes exist to replicate themselves, but if they find themselves on linear DNA, every replication brings them one step closer to non-existence. Viewed this way, it is a wonder our genes survive embryonic development, because we all start out with a single set of linear chromosomes that are replicated countless times before we are born. So how on earth do we still have any DNA?

This mystery was solved in the early 1970s with the discovery of an enzyme known as *telomerase*. In embryos, telomerase is constantly busy elongating *telomeres*, the repetitive, gene-less regions found at the ends of our chromosomes. Thus, although the countless replications occurring in embryonic DNA do shorten an embryo's telomeres, telomerase elongates them again to protect the genes further away from the chromosome ends.[15] Problem solved, right? Well, not exactly. When we are newborns, most of our cells stop producing telomerase,[16] and each subsequent replication chips away at our telomeres. Eventually, when our telomeres have been shortened into erasure, DNA replication can no longer take place without shortening the genes located at the ends of our chromosomes. At this point, our cells stop dividing because they can no longer safely replicate their genes. And that is why aging is associated with a decline in cell divisions.[17]

One possible solution to this problem is to revive telomerase, so to speak. That's what cancer cells do. Cancerous tumors can grow at a devastating rate because they are composed of cells that divide very rapidly. Normally, this would result in accelerated telomere shortening and a halt to cell divisions, but cancer cells sidestep this hurdle by reactivating telomerase production.[18] This, however, is a rather extreme solution to the problem of telomere erosion. An alternative solution is to forgo telomeres and just organize genes in *circular* DNA molecules. Although circular DNA is traditionally associated with bacteria, it can be found in a range of species. Humans, for instance, form extrachromosomal circular DNA in association with pregnancy[19] and various diseases.[20] And *kinetoplastids*, a class of intriguing unicellular organisms, have evolved the awe-inspiring *kinetoplast*, a giant network comprised of thousands of interlocked DNA *minicircles*. Circular DNA replication, it turns out, is so much simpler than linear DNA replication, that even the chain-mail-like kinetoplast can pull off replication with ease. To do this, it gradually releases minicircles to undergo replication at two antipodal sites flanking the kinetoplast.

The newly formed minicircles are then attached to the kine-toplast's periphery as the whole network spins to ensure their uniform distribution.[21]

But circular DNA replication is not always straightforward. Circular DNA molecules, like necklaces in a jewelry box, are prone to twisting and knotting. This hinders their replication as well as the expression of some of their genes, which cannot be accessed. To solve *this* problem, organisms have evolved *topoisomerases*, or enzymes that untwist DNA using brute force. If you've ever untied a knotted necklace, you probably know that this can be a complex, multistep process. Topoisomerases, it seems, lack the patience for this. Instead of carefully untying DNA, they just cut it and push other segments through the gap before repairing it.[22]

Twisting, of course, does not always spell trouble for DNA. In fact, the double helix of DNA would not be helical if the two strands did not twist around each other. How they twist around each other, though, is significant. Most DNA is *right-handed*,[23] which means that it turns the same way your fingers do when you use your right hand to give someone a thumbs-up. But short segments of DNA can also become temporarily *left-handed*. When this happens, right-to-left junctions form at both ends of the left-handed segment.[24] Unsurprisingly, the strain at these peculiar junctions often breaks the DNA, resulting in an increased risk of mutations.

Left-handed DNA has consequently been linked to diseases such as Lupus, Crohn's disease and various cancers.[25] Moreover, an intriguing study that compared the DNA found in the brains of healthy people, people with moderate Alzheimer's disease, and people with severe Alzheimer's disease revealed that while healthy people exclusively have right-handed DNA in their brains, left-handed DNA can be found in the brains of those suffering from severe Alzheimer's disease. But what about people with moderate Alzheimer's disease? Many of their brains, it turns out, contain DNA in an intermediate conformation.[26]

DNA can get even weirder. Sometimes, the double helix gives way to a triple helix, or *triplex*, as in the case of the frataxin gene mutation we encountered in Chapter 5. This triplex, you may recall, is the root cause of Friedreich's ataxia, a disease characterized by neurological symptoms such as sensory loss, mobility issues and impaired speech. Friedreich's ataxia is not the only disease associated with triplex DNA. Lymphangioleiomyomatosis, a rare lung-destroying disease, and hereditary persistence of fetal hemoglobin, a self-explanatory condition in which fetal hemoglobin production persists well into adulthood, have also been linked to triplex DNA formation.[27] There are, of course, other triplex-associated diseases, but these three suffice to demonstrate how the reorganization of DNA into a triple-helical structure impacts everything from our nervous system and lungs to the blood coursing through our veins.

If you think triplex DNA sounds like the title of a science fiction book, wait until you hear about *quadruplex* DNA. The human genome contains over 700,000 regions that can form quadruple-helical structures.[28] These regions are, unsurprisingly, often hotspots for mutations. Consequently, the occurrence of many cancer-associated mutations may be stimulated by the formation of quadruplex DNA.[29] The discovery that cancer cell genomes contain more quadruplexes than normal cell genomes lends credence to this hypothesis,[30] but more importantly, it underscores that there is a lot more to DNA than genes. From built-in junk to magnificent three-dimensional structures, DNA – the molecule at the heart of life – is predominantly defined by its curious characteristics.

CHAPTER 8

Bountiful Bacteria

Darwin's ideas, published in *On the Origin of Species* in 1859, weren't immediately adopted by the entire scientific community, to put it mildly. In fact, when Darwin was awarded the Royal Society's prestigious Copley Medal five years later, his watershed book was noticeably absent from the citation, indicating the Royal Society's reluctance to endorse it.[1] It may therefore surprise you to learn that the Royal Society was once a bastion of open-mindedness that welcomed an uneducated foreigner with open arms, paving the way for the founding of a new scientific discipline – bacteriology.

Antonie van Leeuwenhoek was born in 1632 to a basket maker and brewer's daughter in the Dutch Republic. He received no higher education and became a successful draper and haberdasher. What in the world is a draper and haberdasher, you ask? Well, drapery and haberdashery consist, respectively, of selling cloth and other items required for sewing, two occupations rendered largely obsolete by the Industrial Revolution. Indeed, Van Leeuwenhoek lived in a world radically different from ours, in which people were wholly unaware of the existence of creatures too small to be perceived by the naked eye. Being a self-respecting draper and haberdasher, Van Leeuwenhoek wished to examine the quality of threads more closely than magnifying lenses allowed. This desire led to his interest in lens-making and the construction of simple microscopes that were nevertheless far superior to those of his contemporaries. Soon Van Leeuwenhoek was using his microscopes to examine everything he could think of, from plants to plaque he scraped off his own teeth. It occurred to him that the sharp, almost painful sensation he felt when eating pepper could be the result of tiny spikes piercing his tongue. To prove this, Van Leeuwenhoek softened pepper in water before examining it under a microscope. Much to his surprise, the pepper was not alone – tiny entities were moving in the water around it. These entities, he realized, were alive. Thus, Van Leeuwenhoek became the first human to observe bacteria.

At the time, scientific research was the purview of high-born, well-educated men who published formal scientific papers in Latin. Van Leeuwenhoek's 1680 election to the Royal Society for his letters in colloquial Dutch was therefore nothing short of extraordinary. So earth-shattering were his discoveries, in fact, that seventeenth-century celebrities such as Peter the Great and Leibniz flocked to him for a glimpse of his *animalcules*, or 'little animals'. But much to their disappointment, Van Leeuwenhoek never revealed his best microscopes, choosing instead to provide his guests with a mediocre peek at bacterial life.

Fortunately, we now know a great deal about bacteria. Bacteria are *unicellular* creatures, which means that each bacterium is composed of a single cell. Up until just over 2 billion years ago, bacteria were the only life forms on Earth, and they may well have been the earliest life forms to venture onto land.[2] Bacteria come in a variety of shapes and sizes. Most are spherical or rod-shaped, but some are curved, wavy, corkscrew-shaped or star-shaped. *Mycoplasma genitalium*, the smallest known bacterium, measures roughly 300 nanometers, while *Thiomargarita magnifica*, the largest known bacterium, can reach an astonishing length of 2 centimeters. Most bacteria reproduce by splitting into two cells, but some, like the colossal *Epulopiscium viviparus*, produce as many as 12 intracellular offspring which are released in a birth-like manner.[3] Bacterial cells also come with a variety of molecular add-ons. Some bacteria sport *flagella* with rotary motors, which propel them. Some are covered in tiny, hair-like *pili*, which play a role in both motility and interactions with the environment. Others appear to overdo it by rocking both flagella *and* pili.

Bacteria, as we've seen, are a diverse bunch. They're also everywhere, inhabiting everything from the air around us to the Earth's crust deep below the seabed.[4] It is estimated that there are as many as 5 *nonillion*, or 5 followed by 30 zeros, bacteria in our ecosystem. But what, exactly, do all these bacteria *do*? Van Leeuwenhoek believed that bacteria exist to provide evidence of God's greatness. He came to this conclusion because their minuscule perfection can be interpreted as proof of a creator's omnipotence. Little did he know that bacteria are much more than passive marvels – they have been shaping society since the dawn of civilization.

* * *

Bacteria have acted as shadowy puppeteers throughout history. Consequently, compiling a list of the many ways in which they've influenced humanity would take an infinitely long time. Let us instead focus on the tale of how a certain bacterium has been transforming societal hierarchies since the sixth century CE. Our story begins with the Roman Empire, which was, in many ways, a victim of its own success. Having expanded into one of the largest empires in antiquity, it ended up being too large to govern effectively. This led to its division into the Western Roman Empire and the Eastern Roman Empire in 395 CE. The Western Roman Empire succumbed to barbarian invasions shortly thereafter, in 476 CE, and was survived by the flourishing Eastern Roman Empire. Emperor Justinian, who came to power in 527 CE, sought to restore the glory of the mighty Roman Empire. To that end, he reconquered Italy, parts of Spain and a great deal of northern Africa. Everything seemed to be going according to plan, and Justinian may well have succeeded in reuniting the great Roman Empire, had it not been for a pesky little bacterium known as *Yersinia pestis*.

In 542 the Justinian Plague hit Constantinople, the capital of the Eastern Roman Empire. Caused by the deadly *Y. pestis* bacterium, the Justinian Plague wiped out up to 50% of the Empire's population – roughly 100 million people![5] The resulting plummeting labor supply took an enormous toll on the Empire's military prowess. Masses of Roman soldiers died, and there simply weren't enough young men to replace them. Not that the Empire could afford to, anyway – the labor shortage also yielded an economic crisis. Needless to say, Justinian's dream was another casualty of the plague. His conquests were swiftly reversed, and the Empire's southern provinces were snatched away by Arab forces. Eventually, the Roman Empire collapsed, taking Western antiquity with it.

When *Y. pestis* pushed the Roman world down the path to annihilation, it also dramatically transformed society. The Roman economy, you see, had traditionally been based on slavery. This, of course, could be said of all Mediterranean societies of that era, but the Roman Empire really outdid itself when it came to slavery. In fact, it is estimated that around a third of the first-century inhabitants of the city of Rome were slaves![6] But when the Justinian Plague created a severe shortage of slave labor, landowners had to get creative. To ensure that their fields remained productive, they were forced to grant plots of land to free laborers who paid their lords a variety of taxes and tithes. *Y. pestis* is therefore responsible for much more than a plague; it single-handedly did away with the slavery of the Roman Empire and established feudalism, the defining societal system of the medieval world.

As we've seen, a mere bacterium felled an empire and reinvented society in the sixth century. But *Y. pestis* wasn't done. Eight centuries later, it made a comeback with history's most infamous pandemic – the Black Death.[7] It is estimated that roughly 100 million people lived in Europe, northern Africa and the Near East in 1346. Within seven years, the Black Death had slashed this number to around 75 million.[8] So drastic was this reduction that huge swathes of farmland were abandoned and reforested. Consequently, millions of new trees began to absorb CO_2, cooling the climate significantly enough to trigger the Little Ice Age.[9] Of course, climate change was just one of the results of the Black Death. The labor supply plummeted once again, and surviving peasants began demanding higher wages, leading to the decline of the feudal system. Spurred by new economic opportunities, many peasants moved to towns and specialized in various crafts and trades. The most successful of these became a newly minted *middle class* in Europe's rapidly growing cities. Thus, *Y. pestis* upended the stark inequality of feudalism and planted the seeds of modern socioeconomic systems.

It is clear we must thank our bacterial saviors for (largely) abolishing slavery and serfdom. But are we indebted to them in any other way? To answer this question, we turn to the early years of World War I. Back then, a crippling shortage of acetone threatened to doom Britain to defeat. Acetone, you see, was required to manufacture the smokeless gunpowder used to fire artillery shells. Before the war, Britain had imported acetone from Germany and Austria, but both of these countries were now fighting Britain, resulting in a skyrocketing demand for acetone. Why not produce acetone in Britain, you ask? At the time, a ton of acetone was produced in a highly inefficient process that required roughly 100 tons of wood. Remove Germany's sprawling forests from the equation, and the feasibility of mass-producing acetone evaporated like, well, acetone.

Fortunately, Britain had a trick up its sleeve. Dr. Chaim Weizmann, a biochemist from the University of Manchester, had recently isolated a bacterium called *Clostridium acetobutyli-cum* that could ferment starch into acetone. Weizmann subsequently developed the *Weizmann process*, a biochemical process that leveraged *C. acetobutylicum* to produce 12 tons of acetone from roughly 100 tons of corn. With Churchill's support, huge industrial plants were constructed to carry out the Weizmann process in Britain, Canada and the USA. And that is the story of how the diminutive *C. acetobutylicum* saved Britain, and perhaps the Allied Powers, from defeat in World War I.

Surprisingly, radically transforming long-lasting social structures and contributing to the winning of World War I are hardly the most impressive of bacterial achievements. Bacteria are also adept at impacting our lives from the lowliest of locations – the Earth's soil. We now know that bacteria are the most abundant microbes found there, with up to 1 billion of them inhabiting a single teaspoon of soil. But what are they doing there? Many bacteria, it turns out, are busy decomposing organic matter. This may not sound particularly glamorous, but it is absolutely critical to our existence. Think of it this way – living beings have been dying since life arose almost 4 billion years ago. Were it not for the bacteria that break down dead organisms, our poor planet would be covered with ever-growing mountains of intact corpses. The same bacteria also decompose the waste all organisms secrete, preventing our environment from becoming a colossal cesspool.

The last two sentences, I admit, have been exaggerated to emphasize the debt of gratitude we owe the bacteria that decompose organic matter. The truth is that, in the absence of these hard-working bacteria, dead organisms and waste wouldn't pile up forever, as this would require organisms to keep reproducing. But organisms couldn't possibly survive, let alone reproduce, without decomposers, due to there being only finite quantities of key elements in our environment. Put simply, there is only so much carbon, phosphorus, nitrogen, etc. around us with which we can build and feed our cells. Bacteria don't just break down dead organic matter; they also return valuable elements to the environment, where they are reused by living organisms in a never-ending cycle. Decomposing bacteria can therefore also be thought of as avid recyclers who labor incessantly to save us all from extinction.

* * *

As we have seen, the bacteria around us are, and have always been, immensely influential. But the sway that they hold over the natural world goes deeper than just shaping our environment – bacteria are also tremendously influential *within* our bodies. This may strike you as odd, given that bacteria are commonly considered unwanted invaders our bodies strive to keep out. The truth, however, is that from butterflies[10] to baboons,[11] many animals happily host diverse bacterial populations. Ruminants are a famous example of this phenomenon. Have you ever wondered why cows can digest grass, whereas humans, who share 80% of their genes, cannot? The answer is that the cows themselves are incapable of digesting grass, but the multitudes of bacteria living in their digestive systems decompose the grass for them. This bacterial fermentation, incidentally, produces methane as a by-product. Methane, which is over 25 times more effective than CO_2 at trapping heat in the atmosphere, is subsequently burped out by cows. Thus, bacteria residing within cows around the world emit hundreds of billions of pounds of a perilous greenhouse gas into the atmosphere every year.

The digestive systems of cows are hardly alone in the bacterial prime real estate market. In humans, bacteria flock everywhere from bladders and belly buttons to breast milk, though most of them reside in our large intestine. It is estimated that around 39 trillion bacteria live in and on an adult male's body, a number exceeding that of his own cells by a measly 9 trillion.[12] In other words, we're all technically human–bacterial hybrids. In fact, we were only ever truly human inside our mothers' wombs, because humans typically ingest their first bacteria as they pass through the birth canal.[13]

Having reclassified ourselves as human–bacterial hybrids, it is time we turned our attention to what bacteria do in our shared bodies. Do they play any important roles? Or are they useless hitchhikers we happened to pick up at birth? A first clue comes from some funny-looking amphibians called *caecilians*. Although they are closely related to salamanders, caecilians lack limbs and look like giant worms. Some of these weird creatures also engage in a particularly bizarre type of feeding known as *skin-feeding*. This consists of the caecilian mothers developing a layer of fatty skin which their young tear off with specialized teeth. When caecilian offspring consume their mother's skin, they ingest more than just nutrients – mouthfuls of maternal bacteria are also passed down via this route.[14] And the fact that multiple methods have evolved to supply young with bacteria suggests that bacteria are somehow vital to our survival.

Let us return to the bacteria present in human bodies. Recall that most of these can be found in our large intestine, where they are absolutely essential. Many of these bacteria are busy breaking down food, regulating the absorption of nutrients, and even synthesizing vitamins such as vitamin K.[15] Given their location, it is easy to assume that is all they do, but this couldn't be further from the truth. In fact, bacteria modulate hormone-secreting cells in the intestinal wall and release substances into the bloodstream, extending their influence far beyond the digestive system.

The immune system, for example, is inextricably linked to intestinal bacteria. In one study designed to demonstrate this link, laboratory mice were given antibiotics to kill off their intestinal bacteria. The result? A weakened immune response to viral infection![16] A second noteworthy study found that enrichment of the intestinal bacterium *Enterococcus faecalis* is a reliable predictor of COVID-19 disease severity.[17] In other words, the composition of our intestinal bacterial population determines our ability to fight off diseases.

Obesity is also squarely inside the bacterial sphere of influence. Why are some people more likely than others to become obese? The root cause of this phenomenon is famously hard to pin down. Is it genetic? Environmental? A little bit of both? Twins, with their similar genetic make-up and environmental conditions, are particularly useful to obesity researchers because if only one twin is obese, identifying differences between the two may highlight obesity-promoting factors. That is why an eye-opening study analyzed the intestinal bacterial composition of 77 obese people and their non-obese twins. The results, which may discourage any dieting readers, were nothing short of astounding. The types of intestinal bacteria in obese subjects were markedly different from those found in their non-obese twins, demonstrating that bacteria determine the way our body stores fat. Moreover, the intestines of obese subjects were found to possess lower bacterial diversity,[18] proving that it really does take all sorts!

But wait, what if we got it all backwards? What if obesity in fact causes certain types of intestinal bacteria to thrive at the expense of others, resulting in the observed differences in bacterial composition and diversity? This is a valid question, but a difficult one to address ethically in experiments with human subjects. Fortunately, many obesity researchers work with laboratory mice. But before we delve into insights gleaned from obese mice, two preliminary facts must be presented. The first is that laboratory mice have the unsavory habit of ingesting feces. This means that if two mice are placed in the same cage, they will inevitably consume each other's droppings, effectively swapping intestinal bacteria. The second is that it is possible to raise bacteria-free laboratory mice. This is useful because before scientists started raising bacteria-free mice, any experimental result could have been attributed to variable bacterial compositions. Bacteria-free mice solved this problem by allowing researchers to either eliminate bacterial influences from their experiments or choose the bacteria they wish to introduce into their subjects' intestines. If they're interested in the latter, they can – you guessed it – put bacteria-free mice in the same cage with mice possessing the desired intestinal bacteria.

Armed with this repulsive information, let us explore an intriguing experiment designed to shed light on the development of obesity. In this experiment, a group of bacteria-free mice was divided into two subgroups. One subgroup was placed in a cage with obese mice, and the other subgroup was placed in a cage with lean mice. Naturally, the bacteria-free mice in both cages, ahem, acquired their roommates' intestinal bacteria. But although both cages were supplied with identical feed, the mice placed in a cage with obese mice became obese, whereas the mice placed in a cage with lean mice remained lean![19] We can therefore safely say that certain intestinal bacterial compositions cause obesity – not the other way around.

Are insights derived from laboratory mice applicable to humans? Not always. But an additional study concluded that the bacteria–obesity link discussed above exists in humans, as well. In this study, obese humans and their non-obese twins provided stool samples, which were fed to bacteria-free mice on an otherwise identical diet. As expected, the mice that received bacteria from obese humans became obese, while the mice that received bacteria from non-obese humans remained lean. And just like obese humans, the obese mice were shown to possess reduced intestinal bacterial diversity.[20]

Avoiding calorie-rich food and staying physically active are commonly considered vital to staving off weight gain. And you should definitely do both, for myriad reasons! But when it comes to our figures, it appears that diet and exercise can only do so much. Readers who remain unconvinced should ask themselves why individuals with an abundance of *Akkermansia muciniphila* bacteria are highly unlikely to develop obesity.[21] Or why, for that matter, the ratio of bacteria known as Firmicutes and Bacteroidetes can be used to accurately predict an individual's propensity for obesity.[22] Intestinal bacteria, it is clear, play a key role in determining our weight.

Obesity is just the tip of the iceberg – intestinal bacteria also modulate our brain chemistry to tweak our values. To demonstrate this surprising phenomenon, researchers instructed a group of study participants to play the 'ultimatum game', in which one player decides how to split a given sum of money with a second player. The second player may accept or decline the first player's offer, but if the offer is declined, neither player receives any money. The ultimatum game is fascinating because it is in the first player's best interests to keep most of the money, just as it is in the second player's best interests to accept whatever offer the first player makes. This means that if both players were completely rational, the first player would always offer the second player the smallest possible fraction of the sum, and the second player would always take it because, well, it's better than nothing. But in practice, second players tend to respond to such flagrant unfairness by rejecting the offer, happily forgoing the money they could have pocketed to punish shameless first players. First players, anticipating this reaction, typically make more balanced offers, increasing their chances of walking away with some of the money.

Having completed a round of the ultimatum game, the researchers divided the study participants into two groups. One group took dietary supplements that contained *probiotics*, or beneficial bacteria, and *prebiotics*, or nutrients that promote intestinal colonization, while the other group took a placebo. Seven weeks later, the study participants played another round of the ultimatum game, but this time, members of the group that took probiotics were much more likely to reject unfair offers – even if they were only slightly unfair! Moreover, participants who had a more unbalanced ratio of Firmicutes and Bacteroidetes before taking probiotics developed the highest sensitivity to fairness. Members of the placebo group, on the other hand, played much as they had played in the first round, indicating that our intestinal bacteria shape our perception of fairness.[23]

If you find this hard to believe, rest assured that intestinal bacteria have long been known to influence the brain. Bacteria-free rodents, for example, are more likely to be hyperactive,[24] anxious,[25] and unsociable.[26] But how can intestinal bacteria (or a lack thereof) impact our brains in such life-altering ways? To answer this question, we must first understand how the brain works. The human brain contains around 86 billion cells known as *neurons*,[27] which communicate with each other by secreting and absorbing a variety of chemicals known as *neurotransmitters*. Consequently, neurotransmitters determine everything from muscle movement to mental health. And here's the kicker – the vast majority of intestinal bacteria secrete neurotransmitters that can make their way up to the brain! This explains why people suffering from depression have been shown to lack two bacterial genera, *Coprococcus* and *Dialister*,[28] and why an abundance of a third genus, *Eggerthella*, is associated with depression, anxiety, bipolar disorder and psychosis.[29]

But that's not all – certain bacteria, as it turns out, are associated with schizophrenia. A study that analyzed the stool samples of schizophrenic individuals revealed very specific differences between the intestinal bacterial composition of schizophrenics and healthy people. So specific are these differences, in fact, that it is possible to forgo psychiatric evaluations and identify people suffering from schizophrenia based on their stool samples. Needless to say, when bacteria-free mice were fed stool samples from schizophrenic patients, they developed schizophrenia-like symptoms.[30]

Differences in intestinal bacteria may also explain why some of us struggle with addiction. One fascinating study on alcohol addiction allowed rats to press a lever and be immediately rewarded with alcohol. After a while, the researchers extended the time it took to receive alcohol after pressing the lever. Some rats didn't seem to mind the delay, but others pressed the lever repeatedly and frantically. The researchers then introduced a lever that gave rats who pressed it an uncomfortable shock. This deterred most rats from seeking alcohol, but some persisted in pressing the lever, suggesting that they couldn't help themselves. The intestinal bacterial composition of these rats was subsequently compared to that of the rats who didn't display signs of addiction, and differences in key bacterial species emerged.[31] This, of course, suggests that certain intestinal bacterial compositions may predispose us to addiction.

Last but not least, bacteria are also common in a variety of cancerous tumors, where they can be found in and around both cancer cells *and* immune cells! Interestingly, each cancer type is characterized by a distinctive bacterial composition. This means that, hypothetically speaking, if an unlucky patient were to suffer from both a brain tumor and a breast tumor, the bacterial compositions of the two tumors would differ greatly. However, the bacterial composition of the brain tumor would resemble that of another patient's brain tumor, and the same would be true for the breast tumor and another patient's breast tumor.

What, exactly, are bacteria doing in tumors? We know that the bacterial compositions of tumors correlate with different responses to immunotherapy,[32] so some bacteria must be on the tumor's 'side'. This seems to be true for bacteria in oral and colorectal tumors, which have been shown to suppress the immune system and promote the spread of cancer cells.[33] Moreover, certain anti-cancer drugs may actually work by killing the bacteria in tumors, suggesting that bacteria are vital to tumor development.[34] Given all these remarkable findings, it can be said with certainty that our lives are wholly dictated by our bountiful bacteria.

CHAPTER 9

Flexible Fungi

One of the most intriguing specimens collected by Darwin was an orange golf ball-like fungus from the southern tip of South America. Darwin wrote that this organism, which was later named *Cyttaria darwinii* in his honor, possessed a 'slightly sweet mucous taste', which may have sounded appetizing at the time. After studying it, he reached the conclusion that it was a 'cryptogamic plant', reflecting the erroneous belief of his time that fungi are plant-like organisms.[1]

Wait, aren't fungi plants? Not even close. Fungi lack roots, stems and leaves, rendering them fundamentally different from plants. Indeed, their common conflation with plants seems downright ludicrous when one considers that fungi dominated the planet for roughly 600 million years before the ascension of plants.[2] Moreover, when the algal ancestors of all land plants first moved out of the water, they lacked roots and required fungal help to extract nutrients from the soil. Old habits die hard, and when roots evolved, nutrient-extracting fungi became their vital inhabitants. The 374,000 or so plant species in the world,[3] upon which all terrestrial food chains depend, are therefore wholly dependent on fungi.

Having established that fungi are most certainly not plants, we turn to the question of what fungi *are*. The answer, as it turns out, is far from straightforward. Although there are almost 4 million fungal species,[4] they differ greatly, to put it mildly. Some species of fungi, such as the yeast used to bake bread, are unicellular, meaning that each fungus is a single cell. Others are networks of many *hyphae*, or microscopic tubular structures that branch out on and in food sources, absorbing nutrients with fungal fervor. A third, *dimorphic* group of fungi can exist as either single cells or hyphal networks. This is, surprisingly, one of the *least* confusing things about fungi. Far more bewildering is the way fungi force us to rethink the concept of individual organisms.

Let's start at the beginning: what is an individual? To answer this question, it can be helpful to ponder how you define yourself in contrast to others. Upon reflection, you may reach the conclusion that you are characterized by unique anatomical parts, such as a prominent nose and an irritable bowel. But fungal parts often lack this clear-cut definition. *Holocarpic* fungi, for example, can turn into one big reproductive organ, rendering the concept of anatomical parts, and the variation thereof, meaningless.

Perhaps, then, it is not the characteristics of your organs that define you, but rather the way they come together to form your body. It is, after all, all too easy to determine where your body ends and another individual's body begins. But fungi can lack such salient limits. To understand how this can be true, let us delve deeper into hyphal reality. As we've seen, many fungi are tangled networks of microscopic tubes known as hyphae. It would be difficult to overstate just how prevalent hyphae are. In fact, if the hyphae in a single gram of soil were laid end to end, their length would exceed six miles[5] – the equivalent of approximately 193 Olympic swimming pools! How do hyphae pull off this astonishing growth? Each hypha is a pressurized tube filled with fluid that extends at the tip. The materials required to support this extraordinary extension are packaged into little containers throughout the hypha and rapidly transported by a system of protein motors to a structure at the tip called the *Spitzenkörper*. The Spitzenkörper, which functions as a sorting center for hyphal building materials, dissolves when hyphal extension ceases and reappears when it resumes.[6] Thanks to its tireless efforts, hyphae can extend themselves indefinitely, resulting in limitless fungi. But that's not all – underground hyphae also link plants together in a massive network wittily nicknamed the *Wood Wide Web*, which allows plants to exchange water, valuable elements and vital information.[7] Needless to say, fusion with other organisms further confounds attempts to determine where a fungus ends.

If we can't determine where one fungus ends and another fungus begins, perhaps we are focusing on the wrong dimension. After all, you also have limits in the fourth dimension – time. This is just an overly complicated way of stating that every human's life begins and ends at specific times, and that these temporal limits can be used to define an individual. You probably see where I'm going with this – some fungi can't be defined by temporal limits because they're immortal! Take, for example, the 'Humongous Fungus' of Oregon's Malheur National Forest. This fungal behemoth covers an area of roughly three and a half square miles, and has been alive for almost 10,000 years,[8] proving that death is far from mandatory for fungi. Moreover, although many unicellular creatures age and die, unicellular fungi have been shown to buck this trend. When yeast cells, for instance, are grown under favorable conditions, they forget to age and skip dying altogether.[9]

There are, of course, other ways to define individuals. One could argue that it is your brain that defines you, given that it is the organ where you store your memories and make your decisions. Use your brain to think about it this way – an individual can lose almost any body part, and still remain the same individual. Why can't even the most extreme changes to our body plan alter our identity? Because as long as we're dealing with the same brain, we consider its owner the same individual. Using brains to define individual fungi, however, is impossible for two reasons. The first is that fungi lack brains. The second and more important reason is that fungi can store memories and make decisions in a completely decentralized manner, which explains why they don't need brains in the first place.

In one fascinating experiment on the subject, some hyphal fungi were temporarily exposed to a growth-inhibiting heat shock, while others were spared this discomfort. When both groups were later exposed to a more severe heat shock, hyphae that had experienced the initial heat shock bounced back faster,[10] suggesting retention of the memory of the previous heat shock and the compensatory mechanisms employed to deal with it. Hyphae only retained this memory for 12 hours, after which the forgetful fungi could once again be caught off guard by elevated temperatures.

Memory retention is by no means restricted to hyphal fungi. In fact, unicellular fungi frequently perform far more impressive feats of memorization. Yeast cells, for example, can remember being exposed to a mild dose of salt – for them a deeply unpleasant experience. If the same yeast cells are later exposed to other kinds of chemical stress, the memory of salt exposure significantly improves the cellular response to these new stressors. But that's not all – yeast cells also pass memories down to their descendants. Consequently, a tolerance to chemical stressors persists a whopping five generations after salt exposure![11]

Unless you're going through a breakup, memories are not all upsetting. The same holds true for fungal memories. In an enlightening experiment designed to uncover more positive memories, a wooden block with a fungus was placed in a soil tray with a food source. At first, hyphae from the wooden block extended themselves in all directions, but once the food source was discovered, the fungus (understandably) rerouted its hyphae in the food source's direction. When the wooden block was subsequently moved to a new soil tray, the hyphae did not extend themselves in all directions, as they did in the first soil tray. Instead, they emerged from the side of the wooden block that had previously faced the food source, indicating that they remembered the food source's spatial location and acted accordingly.[12] Hyphal networks are therefore capable of storing memories and making decisions, and given their previously discussed limitless nature, distinct brains or similar structures cannot be used to define individual fungi.

* * *

Brains may be overrated, that much is true. But surely there must be other defining factors that can help us discern ourselves from our peers. Sex, a core characteristic that dictates how our bodies look and with whom we can mate to produce offspring, seems promising. If one examines a population of, say, echidnas, defining individuals as either males or females is quite simple. All one has to do is look for gendered anatomical features, and if the echidnas' spikes get in the way, one can focus instead on mating compatibility. Fungi, however, are much trickier. For starters, their sexes are called *mating types*, for reasons that will soon become clear. Most fungi are capable of both sexual and asexual reproduction, rendering mating type moot for indefinite periods of time. Moreover, some fungi can switch their mating type. Yeast cells, for instance, have two mating types, 'a' and 'α', that differ by a single gene. This means that a yeast cell of the 'a' mating type has an 'a' gene, while a yeast cell of the 'α' mating type has an 'α' gene. Amazingly, each of these cells also has a copy of the opposite mating-type gene stashed away in its genome. This other copy is normally silenced, but every once in a while, an enzyme will snip and switch the two mating-type genes, allowing the silenced gene to take over.[13] How, then, can yeast cells be defined by mating type?

Some fungi have a *tetrapolar* mating system, and this is where things get really crazy. In these fungi, mating type is determined by the combination of two genes found in two distinct genomic regions – A and B. Now, let us say that A can contain either one of two gene variants, A1 and A2, and that B can contain either one of two gene variants, B1 and B2. The resulting four combinations – A1B1, A1B2, A2B1 and A2B2 – yield four different mating types. But what if we allow for more gene variants? Evolution certainly has. That's how *Schizophyllum commune*, a tropical fungus, ended up with over 339 A-gene variants and 64 B-gene variants, and more than 23,000 mating types![14]

At first glance, this bizarre system seems to facilitate the definition of individual fungi. There are, after all, tens of thousands of unique ways to label them! But what does mating type actually mean in *S. commune*? Not much, as it turns out. The different mating types aren't characterized by any physical differences. And mating compatibility requirements are watered down, allowing fungi of each mating type to mate with almost all others. This is actually more sensible than it sounds. *S. commune* doesn't get out much – it grows on decaying wood, which means that its dating pool is limited to close neighbors. Given these slim pickings, it makes sense to set aside old-fashioned sexism and increase the number of mating types that are compatible with any given fungus. One can only marvel at the irony of *S. commune* evolving countless mating types while simultaneously ensuring that the concept of mating type lost any semblance of meaning.

Mating type, as we've seen, is not a useful individual-defining tool. But exploring its intricacies did reveal something that could potentially be employed to define individual fungi – genetic differences. The specific nature of our genomes appears to be at the heart of individuality. What sets you apart from all other organisms? When a sperm cell first fused with an egg cell, producing the first cell in your body, an additional fusion occurred behind the scenes. The sperm cell's nucleus, containing genes from your father, merged with the egg cell's nucleus, which contained genes from your mother. This created a unique, never-before-seen combination of genes that defines you, and you alone, barring identical twins. When your first cell began to divide, giving rise to a body, this one-of-a-kind set of genes was replicated over and over again. Thus, every one of the trillions of cells that make up your body came to contain a nucleus with your signature gene collection.

If a signature gene collection seems to be an iron-clad identifier of individuals, it is, at this point, unsurprising that fungal evolution has sidestepped this integral feature. Hyphae are usually divided by many *septa*, or partitions, that split them up into many cells. These cells, however, are not isolated. Most septa possess a large, central pore through which neighboring cells can exchange components. Which components, if any, could fungal cells wish to exchange? Nuclei, of course! Fungal nuclei tend to be small and flexible, allowing them to regularly squeeze through septal pores. This may strike you as bizarre, given the usual permanence of nuclei, but it is actually a mandatory stage of fungal mating.

S. commune, for example, mates by fusing a hypha with another fungus's hypha. The partnered hyphae swap nuclei, allowing fertilizing nuclei to enter each fungus. Once inside, the fertilizing nuclei shy away from fusing with resident nuclei, opting instead to zip through the hyphal network at high speeds,[15] dividing as they go. Consequently, the entire network is swiftly colonized by an additional, separate set of nuclei. The distribution of these nuclei is haphazard at best, resulting in cells with no nucleus, cells with one nucleus, cells with two different nuclei, and cells with multiple nuclei of every conceivable combination.[16] The genetic content of cells is therefore an inadequate identifier of individuals, as is everything else. Truth be told, the very concept of individual organisms dissolves when faced with flexible fungi.

CHAPTER 10

Clandestine Cooperation

I t is only fitting that we start this final chapter where it all began, with simple, selfish genes replicating themselves in the primordial soup. To get a sense of what these early genes were like, we can examine a certain type of plant pathogen known as a *viroid*. Like viruses, viroids hijack cells, but they are much simpler than viruses because they lack envelopes and protein-encoding genes.[1] In other words, viroids are just pieces of self-replicating RNA – which is exactly how we envision the first genes. Eventually, the first genes began to encode proteins. Vestiges of this evolutionary stage are not hard to come by; recently discovered viroid-like elements known as *obelisks* can be found in your mouth and gut. Like viroids, obelisks lack envelopes, but they do contain one or two protein-encoding genes organized in a rod-like shape.[2]

At a certain point, some ancient genes began to envelop themselves with rudimentary membranes, yielding the earliest protocells. This compartmentalization shielded genes from the chemical degradation they risked out in the open. Genes surrounded by primitive membranes could also accumulate and maintain stable concentrations of the elements required for replication, unlike cell-free genes, which were forced to make do with whatever drifted their way. Membranes are therefore clearly advantageous, but how on earth did they arise in the first place? This may be counterintuitive, but the appearance of simple molecular containers is actually highly likely. Recall, for instance, the spontaneously assembling Cowpea Chlorotic Mottle Virus envelope. Viral envelopes are, of course, very different from cell membranes, but the earliest membranes were probably nothing like the membranes we know and love. Indeed, cellular life may be the serendipitous result of genes stumbling into a certain kind of droplets known as *chemically active droplets*, which can grow and divide – just like cells![3] Once compartmentalization was achieved, membrane structure and composition could be optimized for cellular stability. This, too, must have been a seamless process, given that membranes have been shown to spontaneously assemble in simple mixtures of organic molecules.[4]

The first membranes may have surrounded single genes, like the envelopes of multipartite viruses. At some point, multiple genes began to inhabit each protocell, which was evidently advantageous, given that we don't see single-gene cells anymore. But what could have possibly persuaded selfish genes to accept roommates? The answer is that while independent genes have to be laser-focused on replication, intracellular cooperation entrusts a few genes with collective replication and frees the rest to specialize in various aspects of survival. Put simply, many of the genes in each cell could suddenly evolve to fill roles ranging from nutrient acquisition to mobility, accelerating the cell's replication and benefiting all of its genes. And once genes started cooperating, there was no turning back. In fact, some groups of genes found cooperation to be so advantageous that they evolved into *supergenes*, or clusters of linked genes that refuse to be inherited separately.[5]

As we've seen, compartmentalization was a major break-through in the history of life. Cells known as *eukaryotes* don't just use membranes to compartmentalize themselves from their environment; they also use membranes to enclose specialized intracellular compartments, or *organelles*. This allows the organelles to maintain the specific internal conditions required for their various functions. The cell nucleus, where our precious DNA is stored and transcribed into RNA, is the largest and best-known organelle. But how did it arise? Recall that when a virus infects a cell, multiple virus factories known as *viroplasms* appear inside it. Intriguingly, most giant viruses, or *giruses*, produce just one humongous viroplasm, which may have pivoted from storing viral DNA to storing cellular DNA as a newly minted nucleus. This may have happened because a girus stole genes from the ancestor of all eukaryotes and stuffed them into its viroplasm, which gradually morphed into a permanent nucleus.[6] According to this hypothesis, our cells are actually viruses disguised as cells – a decidedly uncooperative state of affairs. Alternatively, a girus may have lent a cell it infected a few viroplasm-building genes, which the cell used to construct an early nucleus.[7]

Eukaryotes are also characterized by the mitochondrion, an energy-producing organelle usually found in multiple copies in each cell. At first glance, the mitochondrion appears to be a run-of-the-mill organelle. But careful examination of its structure reveals that unlike the nucleus, the mitochondrion is surrounded by two membranes! Each mitochondrion also contains its own DNA, which is separate from the nuclear DNA. Moreover, unlike the linear DNA found in nuclei, mitochondrial DNA is circular. Why are mitochondria so bizarre? In the 1960s, an American evolutionary biologist named Lynn Margulis hypothesized that the ancestors of eukaryotic cells did not contain mitochondria until they internalized smaller cells that excelled at producing energy. Over time, these early eukaryotes came to rely on their internalized cells for energy, while the internalized cells became so dependent on the eukaryotes for everything else that they gradually lost now obsolete cellular structures and degenerated to organelles. This intimate cooperation explains why each mitochondrion possesses two membranes – the internal membrane is a relic of the internalized cell's membrane, while the external membrane is the membrane with which eukaryotic cells surround organelles. It also explains the existence of mitochondrial DNA, a vestige of the internalized cell's own DNA.

Margulis's hypothesis was largely dismissed as flapdoodle, and the article she wrote describing it was rejected by a jaw-dropping 15 scientific journals before finally being published in 1967.[8] But we now know that Margulis was right, and that the little cells that had renounced their free-living existence to become mitochondria were bacteria. Bacterial DNA is circular, just like mitochondrial DNA, and as it turns out, the sequence of nucleotides in mitochondrial DNA is very similar to the sequence of bacterial DNA and drastically different from that of nuclear DNA. This is extraordinary, because most eukaryotic cells contain multiple mitochondria, suggesting that the origin of eukaryotes is actually more bacterial than eukaryotic! Humans, of course, are eukaryotes, and as we've seen in Chapter 8, our bodies contain more bacterial cells than human cells. Throw in the fact that even human cells are crammed with degenerate bacteria, and we've got a full-blown identity crisis on our hands.

Why the ancestors of mitochondria were internalized by larger cells in the first place remains unclear. They may have been engulfed by hungry cells that failed to digest them properly. It is also possible that they were parasites that maneuvered their way into larger cells to take advantage of them. Either way, the cooperative relationship that was subsequently established proved immensely beneficial to both partners. And incredibly, after the first proto-mitochondrial bacterium was internalized by a large cell around one and a half billion years ago,[9] the same process occurred *again* in the ancestor of plants. Over 1,000 million years ago, a eukaryotic cell internalized a different kind of bacterium that gradually evolved into a chloroplast, the plant cell organelle responsible for photosynthesis.[10] Needless to say, chloroplasts, too, are enveloped by two membranes, and they also contain circular DNA with a distinctly bacterial sequence.

Internalizing a smaller cell that can evolve into an organelle is a serendipitous, but not necessarily rare, event. We know this because an internalized bacterium is currently evolving into a new organelle of *Braarudosphaera bigelowii* algae. *B. bigelowii*, you see, is incapable of 'fixing' nitrogen, or converting atmospheric nitrogen into vital nitrogenous compounds such as ammonia. So approximately 100 million years ago,

B. bigelowii internalized a small nitrogen-fixing bacterium, which has been gradually morphing into a *nitroplast* organelle. Since that fateful internalization, *B. bigelowii* and its proto-nitroplast have merged their metabolic processes and synchronized their replication cycles, indicating a high level of interdependence. But most importantly, the proto-nitroplast has already lost the ability to produce around half of the proteins its free-living ancestor produced, and now relies on its host to supply it with these proteins.[11] This degeneration, of course, mirrors the process proto-mitochondria and proto-chloroplasts underwent as they adapted to intracellular life.

* * *

The evolution of eukaryotes paved the way for life's next great cooperative leap – multicellularity. As we've seen, all organisms are composed of cells, but some, like bacteria and yeast, are unicellular, while others, like humans and hippos, are multicellular. It is clear that the earliest organisms were unicellular, and that multicellularity evolved independently in many lineages.[12] But why were cells so eager to band together and cooperate? Growth potential is one major reason. Cells can only grow so large before it becomes impossible to shuttle molecules from the membrane to every part of the cell in a timely manner. But cells in a multicellular organism can maintain optimal sizes while simultaneously reaching a monstrous collective size. And it is this cooperative growth that enables multicellular organisms to avoid unicellular predators and beat unicellular competitors. Moreover, multicellularity leaves reproduction to a group of designated cells and frees the rest to specialize in everything from milk production to UV light detection, further increasing an organism's chances of survival and propagation.

Multicellularity is extremely beneficial, but the idea that unicellular organisms somehow sensed this and made the leap to specialized cooperation beggars belief. So how did the first multicellular organisms come to be? The key to this mystery lies in a most unremarkable habitat – ponds. Look closely at pond water and unless you need better glasses, you should be able to spot tiny green balls called *volvoxes* rolling around near the surface. Each volvox is a hollow sphere composed of a few thousand algal cells. The volvox's cells are green because they contain chlorophyll, the green pigment responsible for absorbing light in photosynthesis. To absorb light, the volvox must first detect it, which it does with the help of a light-sensitive organelle found in each cell. Cells then synchronize the movement of their flagella to induce the volvox's delightful rolling towards the light. When it is time to reproduce, little spheres form within the parent volvox, turn inside-out, and then burst out in a bizarre matricidal birth.

Humans are complex multicellular organisms that boast over 200 different types of cells, including various kinds of blood, bone and brain cells. Volvoxes, on the other hand, are simple multicellular organisms with just two cell types – reproductive cells and non-reproductive, or *somatic*, cells. The reproductive cells have obviously undergone extensive specialization; they lack the flagella used by somatic cells to propel volvoxes through water, and are over 500 times larger than somatic cells.[13] The somatic cells, intriguingly, are very similar to *Chlamydomonas reinhardtii*, a closely related unicellular organism.[14] This is important because when *C. reinhardtii* cells encounter unfavorable conditions, they form a *colony*, or cluster of cooperating cells. Unicellular green algae often form colonies to boost their chances of survival when faced with predators,[15] nutrient scarcity,[16] or pollution.[17] But colonies are very different from multicellular organisms because their cells are not specialized, meaning that they all perform the same basic cellular processes. Consequently, when favorable conditions are restored, *C. reinhardtii* colonies can break up into individual cells that resume their unicellular lifestyle. Volvox cells, on the other hand, have undergone specialization, which means that they cannot survive if they are separated, since the somatic cells cannot reproduce and the reproductive cells cannot move.

What can we deduce from the existence of closely related colonial and multicellular organisms? The ancestors of volvoxes were clearly *C. reinhardtii*-like unicellular organisms that achieved multicellularity with the help of an intermediate colonial step, which allowed cells to gradually specialize in reproduction or mobility. During this process of specialization, cells lost fundamental capabilities outside their area of expertise and became increasingly interdependent, until they could no longer survive the colony's dissolution. This, of course, produced the first volvox, a simple multicellular organism with just two cell types. All complex multicellular organisms must have started out this way and gone on to evolve a dizzying array of cell types. In fact, that's exactly what the volvox seems to be doing! Recall that each of its somatic cells contains a light-sensitive organelle. Highly developed light-sensitive organelles can only be found in one specific area of the volvox sphere, implying that the cells there may eventually surrender their flagella and devote themselves to vision.

Incidentally, just as viruses may have given us the genes necessary to build a nucleus, they may have also been generous enough to gift us the genes required for multicellular life. We know this thanks to an eye-opening study on key genes that enabled some of the volvox's cousins to evolve multicellularity. These genes include genes involved in cellular specialization, genes that allow cells to stick to each other, and genes that help cells communicate and exchange a variety of compounds. Amazingly, many of the abovementioned genes in green algae were shown to have viral origins,[18] hinting at the central role viruses played in establishing multicellular cooperation.

For all its advantages, multicellularity is inherently risky because it opens the door to cheating. Once specialization occurs, what's to stop a cell from enjoying the fruits of other cells' labor without contributing anything? Cheaters of this sort are rudimentary parasites that can quickly multiply, leading to certain ruin. Fortunately, this problem is solved by building multicellular organisms from closely related cells. Think of it this way: we all started out as a single cell that divided over and over again to build our bodies. But why must this be such a painstaking process? Why don't random cells just get together to produce our bodies? The answer is that if an organism's cells are closely related, they contain the same genes, and will consequently cooperate to pass them on to the next generation. If, on the other hand, a motley crew of cells were to somehow find themselves in the same body, each cell would be motivated to cheat in order to propagate its genes at the expense of the other cells. It should therefore come as no surprise that *C. reinhardtii* cells are wary of forming colonies with strangers. Instead, they grow until they are more than ten times their usual size and then undergo a rapid succession of divisions to produce colony-mates.[19]

Cheating, then, is disincentivized in multicellular organisms, where all cells contain the same set of genes. Instead of seeking out ways to cheat, cells in multicellular organisms cooperate so selflessly that some of them leap at the chance to sacrifice themselves for the greater good. This remarkable cellular suicide, or *apoptosis*, is actually very common. As embryos, many of our cells underwent apoptosis to ensure our proper development. Our hands, for example, started out as fleshy paddles, and were it not for the apoptosis of cells in key positions, we would have never developed distinct fingers. But apoptosis is not limited to our early development. Damaged cells routinely undergo apoptosis[20] in order to make way for new cells, and if a cell senses that it has the potential to become cancerous, it will usually initiate apoptosis to save the rest of the body from cancer. This is one of the reasons why most potentially cancerous cells do not result in tumors. In fact, tumors can only arise when mutations allow cancerous cells to *bypass* apoptosis and revert to their former, selfish ways.[21]

* * *

Similar cells, as we've seen, find it easy to aggregate and cooperate, but what about dissimilar cells? Or even radically different multicellular organisms? Can they benefit from cooperation? The answer is a resounding yes. Cooperation is breathtakingly universal in the biological world. In fact, even viruses, disparaged earlier as the ultimate parasites, often cooperate with their hosts. As we saw in Chapter 3, some viruses regularly integrate their genes into cellular DNA, and consequently, 8% of the human genome is made up of viral sequences that we have been lugging around for thousands of years.[22] Viral genes are clearly benefiting from this situation, but are they nothing more than shameless exploiters? Or are they secretly cooperating with their long-time hosts?

Think about it this way – when a virus integrates itself into a host's genome, it comes to rely on the host to replicate its genes along with the cellular genes. At this point, killing the host is decidedly disincentivized, because if the host goes down, it will take all integrated viruses with it. That is why some lethal viruses will integrate themselves into a host's genome and subsequently render the host immune to disease. A good example of this remarkable virus–host cooperation can be observed in real time in koalas. The Koala Retrovirus (KoRV) causes Koala Immune Deficiency Syndrome (KIDS), an AIDS-like condition that weakens koalas' immune systems and renders them more susceptible to a variety of diseases. At least, that's what happens to KoRV-infected koalas on Kangaroo Island, off the southern coast of Australia. But koalas living in northern mainland Australia do not suffer from KoRV-associated diseases. The reason? KoRV recently integrated itself into their genomes and mutated into a weakened version of its once virulent self. We know this happened recently because the integrated KoRV can still produce viral particles, unlike the ancient viral sequences discussed above.[23] Indeed, a similar integration may be occurring on Kangaroo Island as we speak!

Speaking of koalas, the placentas they lack are a product of virus–host cooperation. The placenta, you see, plays an almost contradictory role. Large quantities of oxygen and nutrients flow through the placenta, but it must simultaneously ensure a full separation of the maternal and fetal bloodstreams to prevent the mother's immune system from attacking the fetus. How do placentas pull off these contradictory tasks? This is where ancient viruses come in. Some viruses have membranes surrounding their envelopes. To infect a cell, these viruses must first fuse their membranes with the cell's membrane, which they do with the help of a protein known as *syncytin*. Around 25 million years ago, one such virus infected a mammal and integrated itself into the mammal's genome.[24] The viral gene encoding the syncytin protein was subsequently tapped for a new mission – fusing cells to form the *syncytiotrophoblast*, which now serves as a barrier between the placenta and the uterine wall.[25]

The placenta also cooperates with the ancient viruses embedded in our DNA to trick the immune system. This mischief is necessary because if a virus makes it through the placenta, it could kill the developing fetus. To prevent that from happening, the placenta fakes a chronic viral infection which keeps the maternal immune system on its toes. But how can a viral infection be faked? Recall that RNA is normally single-stranded in cells, and that double-stranded RNA tips cells off to the presence of viruses. Some of the ancient viral sequences in our genome produce double-stranded RNA molecules in the placenta, duping the maternal immune system into maintaining a mild immune response throughout the pregnancy. The production of these double-stranded RNA molecules is suppressed in other tissues, indicating that the placenta resurrects degenerate viruses to protect us at our most vulnerable.[26]

The placenta is hardly the only organ we owe to our ancient viral roommates. To read this book, you must make extensive use of another organ you wouldn't have without viral help – your brain. Human brains, you see, are composed of tens of billions of cells known as *neurons*. Each neuron contains an *axon*, a long electrical cable that conducts the electrical impulses involved in everything from processing stimuli to accessing memories.

If you're reading this book indoors, look around for an electrical cable. Assuming everything is up to code, that cable should be covered in rubber to provide electric insulation. Similarly, your axons are covered by an insulating layer of *myelin*, a fatty tissue that paved the way for the evolution of neurons with long, thin axons, packed together in what we now call brains. As it turns out, an ancient virus that integrated itself into our genome is responsible for myelin production, and therefore, for the evolution of complex brains.[27]

Benefiting from one's viruses doesn't always necessitate generations-long cooperation. Infection with the Hepatitis G Virus, for instance, has been shown to slow down the development of AIDS in HIV-positive individuals.[28] Viruses are also able to prevent their hosts from developing diabetes.[29] And some herpes viruses can protect their hosts from infection with *Y. pestis*, the bacterium behind the Justinian Plague and the Black Death.[30] This, however, does not mean that viruses cannot cooperate with bacteria, as well. In fact, *bacteriophages*, or viruses that infect bacteria, are instrumental to the evolution of *superbugs*, or bacteria that are resistant to multiple kinds of antibiotics.[31]

Sometimes viruses, bacteria and humans all cooperate to prolong their intertwined lives. A study of Japanese centenarians revealed a tremendous diversity of intestinal bacteria and bacteriophages,[32] highlighting the benefits of complex cooperative networks. How does this widespread cooperation stave off death? Recall that intestinal bacteria play a key role in our bodies' ability to fight infection. A series of experiments conducted at Oxford demonstrated that while the combination of tens of species of intestinal bacteria prevents certain infections, each of those species is, on its own, significantly less effective at preventing them.[33] The longevity of Japanese centenarians may therefore be linked to the immunity they gain from diverse, cooperative bacteria and viruses.

When cooperative relationships involve parasites, they often turn sinister – as in the case of the *Diachasmimorpha longicaudata* wasp, which parasitizes fruit fly larvae. A virus has been shown to infect both *D. longicaudata* and fruit flies, indicating that it uses *D. longicaudata* to hitch a ride to fruit flies. In return, the virus suppresses the fruit fly's immune system, enabling the wasp to exploit its host freely.[34] Bacteria are also adept at weaponizing their cooperation with viruses. Some bacterio-phages produce toxins to which their host is immune. When these toxins are released, they kill off neighboring bacteria, effectively destroying the host bacterium's competition.[35] Speaking of bacteria, remember *H. bacteriophora* from Chapter 2? This parasitic worm, which burrows into a caterpillar and makes it glow red as it is devoured from within, is surpris-ingly cooperative. When it invades a caterpillar, *H. bacteriophora* releases thousands of *Photorhabdus luminescens* bacteria to feast on its tissues. In return, the luminescent bacteria provide the dead caterpillar with its jarring red glow, scaring away predators and allowing the parasitic worm to snack in safety.[36]

* * *

The same kinds of cooperative relationships often exist between a variety of organisms. In Chapter 8 we delved into how bacteria living in bovine digestive systems break down grass. In return, they are provided with a safe, nutrient-rich habitat – a clear win–win. Like cows, termites have the misfortune of basing their diet on something they can't digest. To break down wood, some termites take a leaf from the cows' book and host more capable bacteria in their digestive systems. Other termites opt to outsource wood digestion to fungi. These fascinating insects cultivate elaborate temperature-controlled fungus gardens, and proceed to regurgitate partially digested wood on the fruit of their labor. The fungi then dutifully decompose the wood, producing compost for the termites' consumption.

Fungi, incidentally, are famous for their role in the most epic of cooperative relationships – the lichen. If you've never heard of lichens, head over to the nearest rock or old building. Do you see any splotches on the surface? Those are lichens, living beings that cover up to 8% of the Earth's land surface.[37] But what, exactly, are lichens? In 1867, Swiss botanist Simon Schwendener hypothesized that each lichen is composed of a fungus and an alga. Naturally, this notion was dismissed and ridiculed by most lichenologists, who refused to accept that different organisms could be so closely intertwined. But Schwendener's 'dual hypothesis' prevailed, and in 1877 the German botanist Albert Bernhard Frank coined the term *symbiosis* to describe the cooperative relationship of fungi and algae in lichens. Lichens subsequently became the poster child for symbiotic relationships, which were discovered to be exceedingly common between both closely related and unrelated organisms.

Symbiotic relationships are universal because cooperation allows each partner to benefit from the other's unique capabilities. In lichens, for example, algae photosynthesize and provide their fungal partners with the resulting carbohydrates. The fungi, for their part, protect their algal partners and provide them with water and nutrients extracted from the environment. It is therefore unsurprising that fungi and algae are very eager to cooperate. Remember *Chlamydomonas reinhardtii*, the unicellular alga with colonial tendencies? Simply growing it alongside yeast results in the rapid development of symbiotic relationships. This spontaneous formation of lichens also occurs when other *Chlamydomonas* species and fungal species bump into each other,[38] indicating that cooperation is an evolutionary default setting.

Scientists strive to build simple models to describe nature. Schwendener's dual hypothesis is a good example of this. But evolution cares little for the elegance of his musings. As it turns out, lichens are actually *multi-partner* symbiotic relationships between various types of fungi,[39] various types of algae,[40] protists,[41] bacteria and viruses![42] This groundbreaking discovery suggests that cooperation is an ever-expanding phenomenon. And this is particularly true for humans. Compare, for instance, the range and scale of cooperative activities performed by ants and by humans. Ants, often admired for their flawless cooperation, only cooperate with each other to achieve a limited number of modest goals, such as feeding their young and protecting their nests. But humans routinely cooperate with billions of perfect strangers via complex mechanisms to accomplish amazing feats, such as battling global pandemics and exploring outer space. Widespread cooperation, it seems, is what makes us human.

In Chapter 1 we learned that it is the selfishness of our genes that drives evolution. But these days, no gene is an island. In fact, each gene must coexist with up to tens of thousands of other genes in a cell, and with an infinite number of other genes in its environment. Consequently, cooperating with other genes is, paradoxically, the most selfish thing a gene can do. And that is why if you look closely enough, life is riddled with clandestine cooperation.

Epilogue

An implicit assumption permeates this book. Genes are just segments of DNA – they are inherently passive. So when we speak of the myriad ways genes compete and cooperate, we assume that the existence of a certain gene will always result in the creation of a certain protein, which will determine a certain trait, which, in turn, will either help or hamper survival. But genes are rarely so deterministic. Let us start with the assumption that genes yield proteins. This is often wrong because many genes are *silenced*, which means that the proteins they encode fail to materialize. An exhaustive review of all gene-silencing mechanisms and their delightful intricacies could easily fill a book, and is therefore ill-suited to the final pages of this one. Let us focus, instead, on just one fascinating example of widespread gene silencing.

Recall that female genomes possess two X chromosomes, one maternal and one paternal, while male genomes are characterized by a single maternal X chromosome and a paternal Y chromosome. In practice, this means that each of a female's cells contains two copies of X chromosome genes, whereas a male's cells contain one copy apiece. Interestingly, females do not produce twice as many X chromosome-encoded proteins as males. The reason? Their cells silence the genes on one X chromosome in a process known as *X-inactivation*. Here's how it works. We all start out as a fertilized egg cell that undergoes a succession of divisions to create an embryo. In placental mammals, each of an early female embryo's cells will randomly select an X chromosome and fold it into a dense, thoroughly silenced structure. Consequently, about half of a female placental mammal's cells will end up with a silenced maternal X chromosome and the other half will end up with a silenced paternal X chromosome. Meanwhile, in marsupials such as koalas, the paternal X chromosome is always singled out for silencing in a bizarre display of molecular feminism. Thus, hundreds of genes are prevented from producing the proteins they encode, rendering them incapable of competition or cooperation.

Surprisingly, when a certain gene *is* allowed to produce proteins, it doesn't necessarily produce the same protein over and over again. As we've seen, the genes in our DNA are transcribed into RNA molecules, which are translated into the chains of amino acids we call proteins. But before an RNA molecule undergoes translation, it must be edited in a process known as *splicing*. This involves snipping out segments of RNA and stitching up the gaps. The sequence of the *spliced* RNA molecule is then translated into a sequence of amino acids to yield a protein. Curiously enough, the vast majority of human genes undergo *alternative splicing*,[1] meaning that various combinations of segments can be snipped out of a given RNA molecule. A single gene can therefore encode multiple proteins with different, or even opposing, functions![2] Given this identity crisis, how can genes be said to be competing or cooperating? To further complicate things, some genes can undergo *trans-splicing*, a process that pastes together RNA segments from *different* genes, resulting in peculiar hybrid proteins that upend everything we thought we knew about how genes work.

Most genes, as we've seen, do not encode exactly one protein. But even if they did, their mere existence would not suffice to determine an organism's traits. This is because many traits depend not on the existence of certain proteins, but on their *levels*. In other words, it's not about whether a certain gene produces a protein, but about the number of copies of that protein it produces. There are numerous factors that determine the number of copies of a given protein in a cell. Take, for example, the location where RNA is translated into proteins. Different cellular compartments yield very different protein levels, which means that genes are not given equal opportunities to catch natural selection's eye. Moreover, translating the same RNA molecule in different cellular compartments can result in proteins with different functions,[3] suggesting a fluid gene–trait relationship that is hard to reconcile with notions of gene competition and cooperation.

In this book, we defined proteins as chains of amino acids, but technically speaking, they're *folded* chains of amino acids. Each protein possesses a specific three-dimensional structure, or *fold*, that is dictated by its sequence. This is important because a protein's function is wholly dependent on its fold. Put simply, the gene's sequence determines the protein's sequence, which determines the protein's fold, which determines the protein's function, which determines a trait. Armed with this knowledge, scientists have long assumed that when genes compete and cooperate, they do so through their unique protein folds. But we now know that some proteins are *metamorphic proteins*, which can refold and switch functions at the drop of a hat.[4] If protein function is so inexplicably capricious, how on earth does natural selection sort through genes?

To add insult to injury, traits can also be completely gene-independent. We know that a protein's fold is crucial to its function because when proteins *misfold*, they are invariably dysfunctional. But why would proteins stray from their genetically determined folds in the first place? To understand how this could happen, we must first introduce the *PrP^C* protein. PrP^C is produced by many types of human cells, and is especially abundant in our central nervous system, where it appears to play a role in cell survival and cell adhesion.[5] However, when PrP^C misfolds, it becomes a *prion* – a dysfunctional protein that induces the misfolding of any PrP^C proteins it encounters. Thus, the presence of a single prion can rapidly prionize many proteins, leading to the massive accumulation of prions in brain cells. This, in turn, leads to fatal neurodegenerative conditions such as mad cow disease.

Prions are remarkable proteins. Their discovery challenged the prevailing paradigm of genetics, which held that every biological trait can be traced back to a protein structure dictated by a gene. The potency of prions, as we've seen, is derived from their unique structure, which they adopt after *discarding* the structure dictated by a gene. Disregarding the information encoded in genes, in fact, is what prions are all about! They make a point of transmitting their defining trait directly from protein to protein, skipping right over transcription and translation. And their brazen overruling of genes in a gene-centric world is every bit as revolutionary as Darwin's original epiphany.

References

Chapter 1: Genial Genes

1. Holterhoff, K. (2014). The history and reception of Charles Darwin's hypothesis of pangenesis. *Journal of the History of Biology*, 47, 661–695. https://doi.org/10.1007/s10739-014-9377-0
2. Mendel, G. (1866). Versuche über Plflanzenhybriden. *Verhandlungen des naturforschenden Vereines in Brünn*, Bd. IV für das Jahr 1865, Abhandlungen, 3–47.
3. Fisher, R. A. (1936). Has Mendel's work been rediscovered? *Annals of Science*, 1(2), 115–137. https://doi.org/10.1080/00033793600200111
4. Miescher-Rüsch, F. (1871). *Ueber die chemische Zusammensetzung der Eiterzellen*.
5. Lamm, E., Harman, O., and Veigl, S. J. (2020). Before Watson and Crick in 1953 came Friedrich Miescher in 1869. *Genetics*, 215(2), 291–296. https://doi.org/10.1534/genetics.120.303195
6. Jones, M. E. (1953). Albrecht Kossel, a biographical sketch. *The Yale Journal of Biology and Medicine*, 26(1), 80.
7. Griffith, F. (1928). The significance of pneumococcal types. *Epidemiology & Infection*, 27(2), 113–159. https://doi.org/10.1017/S0022172400031879
8. Avery, O. T., MacLeod, C. M., and McCarty, M. (1944). Induction of transformation by a desoxyribonucleic acid fraction isolated from pneumococcus type III. *Journal of Experimental Medicine*, 79(2), 137–158.
9. Crick, F. H. et al. (1961). General nature of the genetic code for proteins. *Nature*, 192(4809), 1227–1232. https://doi.org/10.1038/1921227a0
10. Corbin, R. W. et al. (2003). Toward a protein profile of *Escherichia coli*: comparison to its transcription profile. *Proceedings of the National Academy of Sciences*, 100(16), 9232–9237. https://doi.org/10.1073/pnas.1533294100
11. Clutton-Brock, T. H. et al. (1999). Selfish sentinels in cooperative mammals. *Science*, 284(5420), 1640–1644. https://doi.org/10.1126/science.284.5420.1640
12. Holman, L., Dreier, S., and d'Ettorre, P. (2010). Selfish strategies and honest signalling: reproductive conflicts in ant queen associations. *Proceedings of the Royal Society B: Biological Sciences*, 277(1690), 2007–2015. https://doi.org/10.1098/rspb.2009.2311
13. Woodburne, M. O. and Zinsmeister, W. J. (1982). Fossil land mammal from Antarctica. *Science*, 218(4569), 284–286. https://doi.org/10.1126/science.218.4569.284

14. Kuhn, T. S. (1962). *The Structure of Scientific Revolutions*. University of Chicago Press: Chicago.
15. Stebbins, R. C. (1949). Speciation in salamanders of the plethodontid genus *Ensatina*. Berkeley, CA: University of California Press. https://doi.org/10.2307/1438380
16. Fradin, H. et al. (2017). Genome architecture and evolution of a uni-chromosomal asexual nematode. *Current Biology*, 27(19), 2928–2939. https://doi.org/10.1016/j.cub.2017.08.038
17. Cristofari, G. (2024). Snapshots of genetic copy-and-paste machinery in action. *Nature*, new and views. https://doi.org/10.1038/d41586-024-00112-w

Chapter 2: Pernicious Parasites

1. Adler, S. (1959). Darwin's illness. *Nature*, 184(4693), 1102–1103. https://doi.org/10.1038/1841102a0
2. Botto-Mahan, C. and Medel, R. (2021). Was Chagas disease responsible for Darwin's illness? The overlooked eco-epidemiological context in Chile. *Revista Chilena de Historia Natural*, 94. https://doi.org/10.1186/s40693-021-00104-4
3. Costales, J. A. (2017). Cell invasion by *Trypanosoma cruzi* and the type I interferon response in American Trypanosomiasis Chagas Disease (pp. 605–627). Elsevier. https://doi.org/10.1016/B978-0-12-801029-7.00027-7
4. Mauël, J. (1996) Intracellular survival of protozoan parasites with special reference to *Leishmania* spp., *Toxoplasma gondii*, and *Trypanosoma cruzi*. *Advances in Parasitolology*, 38, 1–51. https://doi.org/10.1016/S0065-308X(08)60032-9
5. Counihan, N. A., Modak, J. K., and de Koning-Ward, T. F. (2021). How malaria parasites acquire nutrients from their host. *Frontiers in Cell Developmental Biology*, 9, 649184. https://doi.org/10.3389/fcell.2021.649184
6. Tsai, I. J. et al. (2013). The genomes of four tapeworm species reveal adaptations to parasitism. *Nature*, 496(7443), 57–63. https://doi.org/10.1038/nature12031
7. Mishina, T. et al. (2023). Massive horizontal gene transfer and the evolution of nematomorph-driven behavioral manipulation of mantids. *Current Biology*, 33(22), 4988–4994. https://doi.org/10.1016/j.cub.2023.09.052
8. Pavlou, G. et al. (2018). Toxoplasma parasite twisting motion mechanically induces host cell membrane fission to complete invasion within a protective vacuole. *Cell Host & Microbe*, 24(1), 81–96. https://doi.org/10.1016/j.chom.2018.06.003
9. House, P. K., Vyas, A., and Sapolsky, R. (2011). Predator cat odors activate sexual arousal pathways in brains of *Toxoplasma gondii* infected rats. *PLoS ONE*, 6(8), e23277. https://doi.org/10.1371/journal.pone.0023277

10. Meyer, C. J. et al. (2022). Parasitic infection increases risk-taking in a social, intermediate host carnivore. *Communications Biology*, 5(1), 1180. https://doi.org/10.1038/s42003-022-04122-0

11. Montoya, J. G. and Liesenfeld, O. (2004). Toxoplasmosis. *Lancet* (London, England), 363(9425), 1965–1976. https://doi.org/10.1016/S0140-6736(04)16412-X

12. Coccaro, E. F. et al. (2016). *Toxoplasma gondii* infection: relationship with aggression in psychiatric subjects. *The Journal of Clinical Psychiatry*, 77(3), 21105. https://doi.org/10.4088/JCP.14m09621

13. Johnson, S. K. et al. (2018). Risky business: linking *Toxoplasma gondii* infection and entrepreneurship behaviours across individuals and countries. *Proceedings of the Royal Society B: Biological Sciences*, 285(1883), 20180822. https://doi.org/10.1098/rspb.2018.0822   *20 18*

14. Fenton, A. et al. (2011). Parasite-induced warning coloration: a novel form of host manipulation. *Animal Behaviour*, 81(2), 417–422. https://doi.org/10.1016/j.anbehav.2010.11.010

15. Cooper, C. E. and Withers, P. C. (2023). Postural, pilo-erective and evaporative thermal windows of the short-beaked echidna (*Tachyglossus aculeatus*). *Biology Letters*, 19(1), 20220495. https://doi.org/10.1098/rsbl.2022.0495

16. Debenham, J. et al. (2012). Year-long presence of *Eimeria echidnae* and absence of *Eimeria tachyglossi* in captive short-beaked echidnas (*Tachyglossus aculeatus*). *Journal of Parasitology*, 98(3), 543–549. https://doi.org/10.1645/GE-2982.1

17. Peñalver, E. et al. (2017). Ticks parasitised feathered dinosaurs as revealed by Cretaceous amber assemblages. *Nature Communications*, 8(1), 1924. https://doi.org/10.1038/s41467-017-01550-z

18. Global Burden of Disease Collaborative Network. Global Burden of Disease Study 2013 (GBD 2013) Incidence, Prevalence, and Years Lived with Disability 1990–2013. Seattle, United States of America: Institute for Health Metrics and Evaluation (IHME), 2015.

Chapter 3: Vivacious Viruses

1. Mushegian, A. R. (2020). Are there 1031 virus particles on earth, or more, or fewer? *Journal of Bacteriology*, 202(9), e00052-20. https://doi.org/10.1128/JB.00052-20

2. Rybicki, E. (1990). The classification of organisms at the edge of life or problems with virus systematics. *South African Journal of Science*, 86(4), 182. https://doi.org/10.1080/10137548.1990.9688004

3. Ho, B., Baryshnikova, A., and Brown, G. W. (2018). Unification of protein abundance datasets yields a quantitative *Saccharomyces cerevisiae* proteome. *Cell Systems*, 6(2), 192–205. https://doi.org/10.1016/j.cels.2017.12.004

4. Sicard, A. et al. (2019). A multicellular way of life for a multipartite virus. *Elife*, 8, e43599. https://doi.org/10.7554/eLife.43599.018

5. Deleris, A. et al. (2006). Hierarchical action and inhibition of plant Dicer-like proteins in antiviral defense. *Science*, 313(5783), 68–71. https://doi.org/10.1126/science.1128214

6. Chen, Y. et al. (2014). The capsid protein p38 of turnip crinkle virus is associated with the suppression of cucumber mosaic virus in *Arabidopsis thaliana* co-infected with cucumber mosaic virus and turnip crinkle virus. *Virology*, 462, 71–80. https://doi.org/10.1016/j.virol.2014.05.031

7. Scola, B. L. et al. (2003). A giant virus in amoebae. *Science*, 299(5615), 2033–2033. https://doi.org/10.1126/science.1081867

8. Sun, S. et al. (2010). Structural studies of the Sputnik virophage. *Journal of Virology*, 84(2), 894–897. https://doi.org/10.1128/JVI.01957-09

9. Liu, Y. et al. (2021). Virus-encoded histone doublets are essential and form nucleosome-like structures. *Cell*, 184(16), 4237–4250. https://doi.org/10.1016/j.cell.2021.06.032

10. Aherfi, S. et al. (2022). Incomplete tricarboxylic acid cycle and proton gradient in Pandoravirus massiliensis: is it still a virus? *The ISME Journal*, 16(3), 695–704. https://doi.org/10.1038/s41396-021-01117-3

11. Garmann, R. F. et al. (2016). Physical principles in the self-assembly of a simple spherical virus. *Accounts of Chemical Research*, 49(1), 48–55. https://doi.org/10.1021/acs.accounts.5b00350

12. Legendre, M. et al. (2014). Thirty-thousand-year-old distant relative of giant icosahedral DNA viruses with a pandoravirus morphology. *Proceedings of the National Academy of Sciences*, 111(11), 4274–4279. https://doi.org/10.1073/pnas.1320670111

Chapter 4: Machiavellian Males

1. Hamlin, K. A. (2021). Darwin's bawdy: the popular, gendered and radical reception of the *Descent of Man* in the US, 1871–1910. *BJHS Themes*, 6, 115–131. https://doi.org/10.1017/bjt.2021.6

2. Godin, J. G. J. and McDonough, H. E. (2003). Predator preference for brightly colored males in the guppy: a viability cost for a sexually selected trait. *Behavioral Ecology*, 14(2), 194–200. https://doi.org/10.1093/beheco/14.2.194

3. Godin, J. G. and Dugatkin, L. A. (1996). Female mating preference for bold males in the guppy, Poecilia reticulata. *Proceedings of the National Academy of Sciences*, 93(19), 10262–10267. https://doi.org/10.1073/pnas.93.19.10262

4. Emlen, D. J. (2008). The evolution of animal weapons. *Annual Review of Ecology, Evolution, and Systematics*, 39, 387–413. https://doi.org/10.1146/annurev.ecolsys.39.110707.173502

5. Hamlin, K. A. (2021). Darwin's bawdy. https://doi.org/10.1017/bjt.2021.6

6. Hespeels, B. et al. (2014). Gateway to genetic exchange? DNA double-strand breaks in the bdelloid rotifer *Adineta vaga* submitted to desiccation. *Journal of Evolutionary Biology*, 27(7), 1334–1345. https://doi.org/10.1111/jeb.12326

7.   Gladyshev, E. A., Meselson, M., and Arkhipova, I. R. (2008). Massive horizontal gene transfer in bdelloid rotifers. *Science*, 320(5880), 1210–1213. https://doi.org/10.1126/science.1156407

8.   Aubin, E., El Baidouri, M., and Panaud, O. (2021). Horizontal gene transfers in plants. *Life*, 11(8), 857. https://doi.org/10.3390/life11080857

9.   Graham, L. A. and Davies, P. L. (2021). Horizontal gene transfer in vertebrates: A fishy tale. *Trends in Genetics*, 37(6), 501–503. https://doi.org/10.1016/j.tig.2021.02.006

10.   Quispe-Huamanquispe, D. G., Gheysen, G., and Kreuze, J. F. (2017). Horizontal gene transfer contributes to plant evolution: the case of Agrobacterium T-DNAs. *Frontiers in Plant Science*, 8, 250007. https://doi.org/10.3389/fpls.2017.02015

11.   Heath, D. J. (1977). Simultaneous hermaphroditism; cost and benefit. *Journal of Theoretical Biology*, 64(2), 363–373. https://doi.org/10.1016/0022-5193(77)90363-0

12.   Warner, R. R. (1975). The adaptive significance of sequential hermaphroditism in animals. *The American Naturalist*, 109(965), 61–82. https://doi.org/10.1086/282974

13.   Valenzuela, N. and Lance, V. (eds). (2004). *Temperature-dependent Sex Determination in Vertebrates* (pp. 1–194). Washington, DC: Smithsonian Books.

14.   Fedder, J. (2023). Sex determination and male differentiation in Southern swordtail fishes: Evaluation from an evolutionary perspective. *Fishes*, 8(8), 407. https://doi.org/10.3390/fishes8080407

15.   Muyle, A. et al. (2021). Epigenetics drive the evolution of sex chromosomes in animals and plants. *Philosophical Transactions of the Royal Society B*, 376(1826), 20200124. https://doi.org/10.1098/rstb.2020.0124

16.   Coelho, S. M. et al. (2018). UV chromosomes and haploid sexual systems. *Trends in Plant Science*, 23(9), 794–807. https://doi.org/10.1016/j.tplants.2018.06.005

17.   Deakin, J. E., Graves, J. A. M., and Rens, W. (2012). The evolution of marsupial and monotreme chromosomes. *Cytogenetic and Genome Research*, 137(2-4), 113–129. https://doi.org/10.1159/000339433

18.   Hvilsom, C. et al. (2012). Extensive X-linked adaptive evolution in central chimpanzees. *Proceedings of the National Academy of Sciences*, 109(6), 2054–2059. https://doi.org/10.1073/pnas.1106877109

19.   Lahn, B. T., Pearson, N. M., and Jegalian, K. (2001). The human Y chromosome, in the light of evolution. *Nature Reviews Genetics*, 2(3), 207–216. https://doi.org/10.1038/35056058

20.   Spatz, A., Borg, C., and Feunteun, J. (2004). X-chromosome genetics and human cancer. *Nature Reviews Cancer*, 4(8), 617–629. https://doi.org/10.1038/nrc1413

21.   Rhie, A. et al. (2023). The complete sequence of a human Y chromosome. *Nature*, 621(7978), 344–354. https://doi.org/10.1038/s41586-023-06457-y

22.   Bellott, D. W. et al. (2014). Mammalian Y chromosomes retain widely expressed dosage-sensitive regulators. *Nature*, 508(7497), 494–499. https://doi.org/10.1038/nature13206

23. Bakloushinskaya, I. and Matveevsky, S. (2018). Unusual ways to lose a Y chromosome and survive with changed autosomes: a story of mole voles *Ellobius* (Mammalia, Rodentia). *OBM Genetics*, 2(3), 1–17. https://doi.org/10.21926/obm.genet.1803023

24. Schartl, M. and Lamatsch, D. K. (2023). How to manage without a Y chromosome. *Proceedings of the National Academy of Sciences*, 120(2), e2218839120. https://doi.org/10.1073/pnas.2218839120

25. Ezaz, T. et al. (2006). Relationships between vertebrate ZW and XY sex chromosome systems. *Current Biology*, 16(17), R736-R743. https://doi.org/10.1016/j.cub.2006.08.021

26. Pennell, M. W., Mank, J. E., and Peichel, C. L. (2018). Transitions in sex determination and sex chromosomes across vertebrate species. *Molecular Ecology*, 27(19), 3950–3963. https://doi.org/10.1111/mec.14540

27. Study reveals how egg cells get so big. (2021, March 4). *MIT News | Massachusetts Institute of Technology*. https://news.mit.edu/2021/study-reveals-how-egg-cells-get-so-big-0304

28. Blackburn, M. W. and Calloway, D. H. (1976). Energy expenditure and consumption of mature, pregnant and lactating women. *Journal of the American Dietetic Association*, 69(1), 29–37. https://doi.org/10.1016/S0002-8223(21)06646-3

29. Ryan, C. P. et al. (2018). Reproduction predicts shorter telomeres and epigenetic age acceleration among young adult women. *Scientific Reports*, 8(1), 11100. https://doi.org/10.1038/s41598-018-29486-4

30. Pollack, A. Z., Rivers, K., and Ahrens, K. A. (2018). Parity associated with telomere length among US reproductive age women. *Human Reproduction*, 33(4), 736–744. https://doi.org/10.1093/humrep/dey024

31. DeChiara, T. M., Robertson, E. J., and Efstratiadis, A. (1991). Parental imprinting of the mouse insulin-like growth factor II gene. *Cell*, 64(4), 849–859. https://doi.org/10.1016/0092-8674(91)90513-X

32. Thomson, A. M., Hytten, F. E., and Billewicz, W. Z. (1970). The energy cost of human lactation. *British Journal of Nutrition*, 24(2), 565–572. https://doi.org/10.1079/BJN19700054

33. Nattrass, S. et al. (2019). Postreproductive killer whale grandmothers improve the survival of their grandoffspring. *Proceedings of the National Academy of Sciences*, 116(52), 26669–26673. https://doi.org/10.1073/pnas.1903844116

Chapter 5: Manipulative Mutations

1. Cavalcanti, A. R. et al. (2004). Coding properties of *Oxytricha trifallax* (*Sterkiella histriomuscorum*) macronuclear chromosomes: analysis of a pilot genome project. *Chromosoma*, 113, 69–76. https://doi.org/10.1007/s00412-004-0295-3

2. Sharma, V. et al. (2018). A genomics approach reveals insights into the importance of gene losses for mammalian adaptations. *Nature Communications*, 9(1), 1215. https://doi.org/10.1038/s41467-018-03667-1

3.   Wiessner, P. W. (2014). Embers of society: Firelight talk among the Ju/'hoansi Bushmen. *Proceedings of the National Academy of Sciences*, 111(39), 14027–14035. https://doi.org/10.1073/pnas.1404212111
4.   Wrangham, R. W. and Carmody, R. N. (2010). Human adaptation to the control of fire. *Evolutionary Anthropology*. https://doi.org/10.1002/evan.20275
5.   Hubbard, T. D. et al. (2016). Divergent Ah receptor ligand selectivity during hominin evolution. *Molecular Biology and Evolution*, 33(10), 2648–2658. https://doi.org/10.1093/molbev/msw143
6.   Swallow, D. M. (2003). Genetics of lactase persistence and lactose intolerance. *Annual Review of Genetics*, 37(1), 197–219. https://doi.org/10.1146/annurev.genet.37.110801.143820
7.   Williams, T. N. et al. (2005). Sickle cell trait and the risk of *Plasmodium falciparum* malaria and other childhood diseases. *The Journal of Infectious Diseases*, 192(1), 178–186. https://doi.org/10.1086/430744
8.   Luzzatto, L. (2012). Sickle cell anaemia and malaria. *Mediterranean Journal of Hematology and Infectious Diseases*, 4(1). https://doi.org/10.4084/mjhid.2012.065
9.   Idnurm, A. et al. (2008). Identification of the sex genes in an early diverged fungus. *Nature*, 451(7175), 193–196. https://doi.org/10.1038/nature06453
10. Rogaev, E. I. et al. (2009). Genotype analysis identifies the cause of the "royal disease". *Science*, 326(5954), 817–817. https://doi.org/10.1126/science.1180660
11. Aumer, D. et al. (2019). A single SNP turns a social honey bee (Apis mellifera) worker into a selfish parasite. *Molecular Biology and Evolution*, 36(3), 516–526. https://doi.org/10.1093/molbev/msy232
12. Bissler, J. J. (2007). Triplex DNA and human disease. *Frontiers in Bioscience*, 12, 4536–4546. https://doi.org/10.2741/2408
13. Hurst, L. D. (2022). Selfish centromeres and the wastefulness of human reproduction. *PLoS Biology*, 20(7), e3001671. https://doi.org/10.1371/journal.pbio.3001671

Chapter 6: Crippling Constraints

1.   Rose, K. D. (2001). The ancestry of whales. *Science*, 293(5538), 2216–2217. https://doi.org/10.1126/science.1065305
2.   Thewissen, J. G. et al. (2007). Whales originated from aquatic artiodactyls in the Eocene epoch of India. *Nature*, 450(7173), 1190–1194. https://doi.org/10.1038/nature06343
3.   Deaner, R. O. et al. (2007). Overall brain size, and not encephalization quotient, best predicts cognitive ability across non-human primates. *Brain, Behavior and Evolution*, 70(2), 115–124. https://doi.org/10.1159/000102973
4.   Aboitiz, F. and Montiel, J. F. (2012). From tetrapods to primates: conserved developmental mechanisms in diverging ecological adaptations. *Progress in Brain Research*, 195, 3–24. https://doi.org/10.1016/B978-0-444-53860-4.00001-5

5. González-Forero, M. and Gardner, A. (2018). Inference of ecological and social drivers of human brain-size evolution. *Nature*, 557(7706), 554–557. https://doi.org/10.1038/s41586-018-0127-x
6. Erecińska, M. and Silver, I. A. (2001). Tissue oxygen tension and brain sensitivity to hypoxia. *Respiration physiology*, 128(3), 263–276. https://doi.org/10.1016/S0034-5687(01)00306-1
7. Lin, M. T. and Beal, M. F. (2003). The oxidative damage theory of aging. *Clinical Neuroscience Research*, 2(5–6), 305–315. https://doi.org/10.1016/S1566-2772(03)00007-0
8. Clavero, M. and García-Berthou, E. (2005). Invasive species are a leading cause of animal extinctions. *Trends in Ecology & Evolution*, 20(3), 110. https://doi.org/10.1016/j.tree.2005.01.003
9. Cahill, A. E. et al. (2013). How does climate change cause extinction? *Proceedings of the Royal Society B: Biological Sciences*, 280(1750), 20121890. https://doi.org/10.1098/rspb.2012.1890
10. Alberts, B. et al. (2002). The initiation and completion of DNA replication in chromosomes. In *Molecular Biology of the Cell*. 4th edition. Garland Science.
11. Bębenek, A. and Ziuzia-Graczyk, I. (2018). Fidelity of DNA replication—a matter of proofreading. *Current Genetics*, 64(5), 985–996. https://doi.org/10.1007/s00294-018-0820-1
12. Brownstein, C. D. et al. (2024). The genomic signatures of evolutionary stasis. *Evolution*, 78(5), 821–834. https://doi.org/10.1093/evolut/qpae028
13. Beavan, A. J., Domingo-Sananes, M. R., and McInerney, J. O. (2024). Contingency, repeatability, and predictability in the evolution of a prokaryotic pangenome. *Proceedings of the National Academy of Sciences*, 121(1), e2304934120. https://doi.org/10.1073/pnas.2304934120
14. McCoy, M. J. and Fire, A. Z. (2024). Parallel gene size and isoform expansion of ancient neuronal genes. *Current Biology*. https://doi.org/10.1016/j.cub.2024.02.021

Chapter 7: Curious Characteristics

1. Eddy, S. R. (2012). The C-value paradox, junk DNA and ENCODE. *Current Biology*, 22(21), R898-R899. https://doi.org/10.1016/j.cub.2012.10.002
2. Park, E. G. et al. (2022). Genomic analyses of non-coding RNAs overlapping transposable elements and its implication to human diseases. *International Journal of Molecular Sciences*, 23(16), 8950. https://doi.org/10.3390/ijms23168950
3. Sweet, A. (2022). Requiem for a Gene: The Problem of Junk DNA for the Molecular Paradigm. MA thesis, University of Chicago. https://doi.org/10.6082/uchicago.5164
4. Shanmugam, A., Nagarajan, A., and Pramanayagam, S. (2017). Non-coding DNA–a brief review. *Journal of Applied Biology and Biotechnology*, 5(5), 42–47.

5.   Romiguier, J. and Roux, C. (2017). Analytical biases associated with GC-content in molecular evolution. *Frontiers in Genetics*, 8, 246001. https://doi.org/10.3389/fgene.2017.00016

6.   Singh, R., Ming, R., and Yu, Q. (2016). Comparative analysis of GC content variations in plant genomes. *Tropical Plant Biology*, 9, 136–149. https://doi.org/10.1007/s12042-016-9165-4

7.   Duret, L. and Galtier, N. (2009). Biased gene conversion and the evolution of mammalian genomic landscapes. *Annual Review of Genomics and Human Genetics*, 10, 285–311. https://doi.org/10.1146/annurev-genom-082908-150001

8.   Kiktev, D. A. et al. (2018). GC content elevates mutation and recombination rates in the yeast Saccharomyces cerevisiae. *Proceedings of the National Academy of Sciences*, 115(30), E7109–E7118. https://doi.org/10.1073/pnas.1807334115

9.   Wu, H. et al. (2012). On the molecular mechanism of GC content variation among eubacterial genomes. *Biology Direct*, 7(1), 1–16. https://doi.org/10.1186/1745-6150-7-2

10. Mugal, C. F., Weber, C. C., and Ellegren, H. (2015). GC-biased gene conversion links the recombination landscape and demography to genomic base composition: GC-biased gene conversion drives genomic base composition across a wide range of species. *BioEssays*, 37(12), 1317-1326. https://doi.org/10.1002/bies.201500058

11. Aliperti, L. et al. (2023). r/K selection of GC content in prokaryotes. *Environmental Microbiology*, 25(12), 3255–3268. https://doi.org/10.1111/1462-2920.16511

12. Picard, M. A. et al. (2023). Transcriptomic, proteomic, and functional consequences of codon usage bias in human cells during heterologous gene expression. *Protein Science*, 32(3), e4576. https://doi.org/10.1002/pro.4576

13. Knight, R. D., Freeland, S. J., and Landweber, L. F. (2001). A simple model based on mutation and selection explains trends in codon and amino-acid usage and GC composition within and across genomes. *Genome Biology*, 2(4), 1–13. https://doi.org/10.1186/gb-2001-2-4-research0010

14. Ohki, R., Tsurimoto, T., and Ishikawa, F. (2001). In vitro reconstitution of the end replication problem. *Molecular and Cellular Biology*, 21(17), 5753–5766. https://doi.org/10.1128/MCB.21.17.5753-5766.2001

15. O'sullivan, R. J. and Karlseder, J. (2010). Telomeres: protecting chromosomes against genome instability. *Nature Reviews Molecular Cell Biology*, 11(3), 171–181. https://doi.org/10.1038/nrm2848

16. Wright, W. E. et al. (1996). Telomerase activity in human germline and embryonic tissues and cells. *Developmental Genetics*, 18(2), 173–179. https://doi.org/10.1002/(SICI)1520-6408(1996)18:2<173::AID-DVG10>3.0.CO;2-3

17. Rossiello, F. et al. (2022). Telomere dysfunction in ageing and age-related diseases. *Nature Cell Biology*, 24(2), 135–147. https://doi.org/10.1038/s41556-022-00842-x

18. Shay, J. W. and Wright, W. E. (2011, December). Role of telomeres and telomerase in cancer. In *Seminars in Cancer Biology* (Vol. 21, No. 6, pp. 349–353). Academic Press. https://doi.org/10.1016/j.semcancer.2011.10.001

19. Sin, S. T. et al. (2020). Identification and characterization of extrachromosomal circular DNA in maternal plasma. *Proceedings of the National Academy of Sciences*, 117(3), 1658–1665. https://doi.org/10.1073/pnas.1914949117

20. Zhao, Y. et al. (2022). Extrachromosomal circular DNA: Current status and future prospects. *Elife*, 11, e81412. https://doi.org/10.7554/eLife.81412

21. Morris, J. C. et al. (2001). Replication of kinetoplast DNA: an update for the new millennium. *International Journal for Parasitology*, 31(5–6), 453–458. https://doi.org/10.1016/S0020-7519(01)00156-4

22. Champoux, J. J. (2001). DNA topoisomerases: structure, function, and mechanism. *Annual Review of Biochemistry*, 70(1), 369–413. https://doi.org/10.1146/annurev.biochem.70.1.369

23. Potaman, V. N. and Sinden, R. R. (2013). DNA: Alternative conformations and biology. In Madame Curie Bioscience Database [Internet]. Landes Bioscience.

24. Ha, S. C. et al. (2005). Crystal structure of a junction between B-DNA and Z-DNA reveals two extruded bases. *Nature*, 437(7062), 1183–1186. https://doi.org/10.1038/nature04088

25. Ravichandran, S., Subramani, V. K., and Kim, K. K. (2019). Z-DNA in the genome: from structure to disease. *Biophysical Reviews*, 11(3), 383–387. https://doi.org/10.1007/s12551-019-00534-1

26. Suram, A. et al. (2002). First evidence to show the topological change of DNA from B-DNA to Z-DNA conformation in the hippocampus of Alzheimer's brain. *Neuromolecular Medicine*, 2, 289–297. https://doi.org/10.1385/NMM:2:3:289

27. Bissler, J. J. (2007). Triplex DNA and human disease. *Frontiers in Bioscience*, 12, 4536–4546. https://doi.org/10.2741/2408

28. Chambers, V. S. et al. (2015). High-throughput sequencing of DNA G-quadruplex structures in the human genome. *Nature Biotechnology*, 33(8), 877–881. https://doi.org/10.1038/nbt.3295

29. Kosiol, N. et al. (2021). G-quadruplexes: A promising target for cancer therapy. *Molecular Cancer*, 20(1), 1–18. https://doi.org/10.1186/s12943-021-01328-4

30. Spiegel, J., Adhikari, S. and Balasubramanian, S. (2020). The structure and function of DNA G-quadruplexes. *Trends in Chemistry*, 2(2), 123–136. https://doi.org/10.1016/j.trechm.2019.07.002

Chapter 8: Bountiful Bacteria

1. Burkhardt, F. H. (2001). Darwin and the Copley Medal. *Proceedings of the American Philosophical Society*, 145(4), 510–518.

2. Homann, M. et al. (2018). Microbial life and biogeochemical cycling on land 3,220 million years ago. *Nature Geoscience*, 11(9), 665–671. https://doi.org/10.1038/s41561-018-0190-9

3. Sannino, D. R. et al. (2023). The exceptional form and function of the giant bacterium Ca. Epulopiscium viviparus revolves around its sodium motive force. *Proceedings of the National Academy of Sciences*, 120(52), e2306160120. https://doi.org/10.1073/pnas.2306160120
4. Lever, M. A. et al. (2013). Evidence for microbial carbon and sulfur cycling in deeply buried ridge flank basalt. *Science*, 339(6125), 1305–1308. https://doi.org/10.1126/science.1229240
5. Eisenberg, M. and Mordechai, L. (2019). The Justinianic Plague: an interdisciplinary review. *Byzantine and Modern Greek Studies*, 43(2), 156–180. https://doi.org/10.1017/byz.2019.10
6. Madden, J. (1996). Slavery in the Roman Empire – numbers and origins. *Classics Ireland*, 3, 109–128. https://doi.org/10.2307/25528294
7. Haensch, S. et al. (2010). Distinct clones of *Yersinia pestis* caused the black death. *PLoS Pathogens*, 6(10), e1001134. https://doi.org/10.1371/journal.ppat.1001134
8. McEvedy, C. (1988). The bubonic plague. *Scientific American*, 258(2), 118–123. https://doi.org/10.1038/scientificamerican0288-118
9. Van Hoof, T. B. et al. (2006). Forest re-growth on medieval farmland after the Black Death pandemic—Implications for atmospheric CO2 levels. *Palaeogeography, Palaeoclimatology, Palaeoecology*, 237(2–4), 396–409. https://doi.org/10.1016/j.palaeo.2005.12.013
10. Phalnikar, K., Kunte, K., and Agashe, D. (2019). Disrupting butterfly caterpillar microbiomes does not impact their survival and development. *Proceedings of the Royal Society B*, 286(1917), 20192438. https://doi.org/10.1098/rspb.2019.2438
11. de Jonge, N. et al. (2022). The gut microbiome of 54 mammalian species. *Frontiers in Microbiology*, 13, 886252. https://doi.org/10.3389/fmicb.2022.886252
12. Sender, R., Fuchs, S., and Milo, R. (2016). Are we really vastly outnumbered? Revisiting the ratio of bacterial to host cells in humans. *Cell*, 164(3), 337–340. https://doi.org/10.1016/j.cell.2016.01.013
13. Blaser, M. J. et al. (2021). Lessons learned from the prenatal microbiome controversy. *Microbiome*, 9(1), 1–7. https://doi.org/10.1186/s40168-020-00946-2
14. Kouete, M. T. et al. (2023). Parental care contributes to vertical transmission of microbes in a skin-feeding and direct-developing caecilian. *Animal Microbiome*, 5(1), 1–15. https://doi.org/10.1186/s42523-023-00243-x
15. Hill, M. J. (1997). Intestinal flora and endogenous vitamin synthesis. *European Journal of Cancer Prevention*, 6(2), S43-S45. https://doi.org/10.1097/00008469-199703001-00009
16. Ichinohe, T. et al. (2011). Microbiota regulates immune defense against respiratory tract influenza A virus infection. *Proceedings of the National Academy of Sciences*, 108(13), 5354–5359. https://doi.org/10.1073/pnas.1019378108
17. Ward, D. V. et al. (2021). The intestinal and oral microbiomes are robust predictors of COVID-19 severity the main predictor of

COVID-19-related fatality. *medRxiv*, 2021-01. https://doi.org/10.1101/2021.01.05.20249061

18. Turnbaugh, P. J. et al. (2009). A core gut microbiome in obese and lean twins. *Nature*, 457(7228), 480–484. https://doi.org/10.1038/nature07540

19. Turnbaugh, P. J. et al. (2006). An obesity-associated gut microbiome with increased capacity for energy harvest. *Nature*, 444(7122), 1027–1031. https://doi.org/10.1038/nature05414

20. Ridaura, V. K. et al. (2013). Gut microbiota from twins discordant for obesity modulate metabolism in mice. *Science*, 341(6150), 1241214. https://doi.org/10.1126/science.1241214

21. Zhou, Q. et al. (2020). Gut bacteria Akkermansia is associated with reduced risk of obesity: evidence from the American Gut Project. *Nutrition & Metabolism*, 17(1), 1–9. https://doi.org/10.1186/s12986-020-00516-1

22. Chakraborti, C. K. (2015). New-found link between microbiota and obesity. *World Journal of Gastrointestinal Pathophysiology*, 6(4), 110. https://doi.org/10.4291/wjgp.v6.i4.110

23. Falkenstein, M. et al. (2024). Impact of the gut microbiome composition on social decision-making. *PNAS Nexus*, 3(5), pgae166. https://doi.org/10.1093/pnasnexus/pgae166

24. Cryan, J. F. and Mazmanian, S. K. (2022). Microbiota–brain axis: Context and causality. *Science*, 376(6596), 938–939. https://doi.org/10.1126/science.abo4442

25. Crumeyrolle-Arias, M. et al. (2014). Absence of the gut microbiota enhances anxiety-like behavior and neuroendocrine response to acute stress in rats. *Psychoneuroendocrinology*, 42, 207–217. https://doi.org/10.1016/j.psyneuen.2014.01.014

26. Sherwin, E. et al. (2019). Microbiota and the social brain. *Science*, 366(6465), eaar2016. https://doi.org/10.1126/science.aar2016

27. Azevedo, F. A. et al. (2009). Equal numbers of neuronal and nonneuronal cells make the human brain an isometrically scaled-up primate brain. *Journal of Comparative Neurology*, 513(5), 532–541. https://doi.org/10.1002/cne.21974

28. Valles-Colomer, M. et al. (2019). The neuroactive potential of the human gut microbiota in quality of life and depression. *Nature Microbiology*, 4(4), 623–632. https://doi.org/10.1038/s41564-018-0337-x

29. Nikolova, V. L. et al. (2021). Perturbations in gut microbiota composition in psychiatric disorders: a review and meta-analysis. *JAMA Psychiatry*, 78(12), 1343–1354. https://doi.org/10.1001/jamapsychiatry.2021.2573

30. Zheng, P. et al. (2019). The gut microbiome from patients with schizophrenia modulates the glutamate-glutamine-GABA cycle and schizophrenia-relevant behaviors in mice. *Science Advances*, 5(2), eaau8317. https://doi.org/10.1126/sciadv.aau8317

31. Jadhav, K. S. et al. (2018). Gut microbiome correlates with altered striatal dopamine receptor expression in a model of compulsive alcohol seeking. *Neuropharmacology*, 141, 249–259. https://doi.org/10.1016/j.neuropharm.2018.08.026

32. Nejman, D. et al. (2020). The human tumor microbiome is composed of tumor type–specific intracellular bacteria. *Science*, 368(6494), 973–980.
33. Galeano Niño, J. L. et al. (2022). Effect of the intratumoral microbiota on spatial and cellular heterogeneity in cancer. *Nature*, 611(7937), 810–817. https://doi.org/10.1038/s41586-022-05435-0
34. LaCourse, K. D. et al. (2022). The cancer chemotherapeutic 5-fluorouracil is a potent Fusobacterium nucleatum inhibitor and its activity is modified by intratumoral microbiota. *Cell Reports*, 41(7). https://doi.org/10.1016/j.celrep.2022.111625

Chapter 9: Flexible Fungi

1. Keynes, R. D. (ed.). (2000). *Charles Darwin's Zoology Notes & Specimen Lists from H.M.S.* Beagle (p. 127). Cambridge University Press.
2. Adamatzky, A. et al. (2022). Fungal States of Minds. *bioRxiv*, 2022-04. https://doi.org/10.1101/2022.04.03.486900
3. Christenhusz, M. J. and Byng, J. W. (2016). The number of known plants species in the world and its annual increase. *Phytotaxa*, 261(3), 201–217. https://doi.org/10.11646/phytotaxa.261.3.1
4. Hawksworth, D. L. and Lücking, R. (2017). Fungal diversity revisited: 2.2 to 3.8 million species. *Microbiology Spectrum*, 5(4), 10-1128. https://doi.org/10.1128/microbiolspec.FUNK-0052-2016
5. Ritz, K. and Young, I. M. (2004). Interactions between soil structure and fungi. *Mycologist*, 18(2), 52–59. https://doi.org/10.1017/S0269915X04002010
6. Reynaga-Peña, C. G., Gierz, G., and Bartnicki-Garcia, S. (1997). Analysis of the role of the Spitzenkörper in fungal morphogenesis by computer simulation of apical branching in *Aspergillus niger*. *Proceedings of the National Academy of Sciences*, 94(17), 9096–9101. https://doi.org/10.1073/pnas.94.17.9096
7. Rhodes, C. J. (2017). The whispering world of plants: 'The wood wide web'. *Science Progress*, 100(3), 331–337. https://doi.org/10.3184/003685017X14968299580423
8. Duke, C. (2021, November 23). The Largest Living Thing on Earth Is a 3.5-Square-Mile Fungus. *Discovery*. https://www.discovery.com/nature/the-largest-living-thing-on-earth-is-a-3-5-square-mile-fungus
9. Coelho, M. et al. (2013). Fission yeast does not age under favorable conditions, but does so after stress. *Current Biology*, 23(19), 1844–1852. https://doi.org/10.1016/j.cub.2013.07.084
10. Andrade-Linares, D. R., Veresoglou, S. D., and Rillig, M. C. (2016). Temperature priming and memory in soil filamentous fungi. *Fungal Ecology*, 21, 10–15. https://doi.org/10.1016/j.funeco.2016.02.002
11. Guan, Q. et al. (2012). Cellular memory of acquired stress resistance in *Saccharomyces cerevisiae*. *Genetics*, 192(2), 495–505. https://doi.org/10.1534/genetics.112.143016
12. Fukasawa, Y., Savoury, M., and Boddy, L. (2020). Ecological memory and relocation decisions in fungal mycelial networks: responses to

quantity and location of new resources. *The ISME Journal*, 14(2), 380–388. https://doi.org/10.1038/s41396-019-0536-3

13. Fraser, J. A. and Heitman, J. (2003). Fungal mating-type loci. *Current Biology*, 13(20), R792-R795. https://doi.org/10.1016/j.cub.2003.09.046

14. Kothe, E. (1999). Mating types and pheromone recognition in the homobasidiomycete *Schizophyllum commune*. *Fungal Genetics and Biology*, 27(2–3), 146–152. https://doi.org/10.1006/fgbi.1999.1129

15. Snider, P. J. and Raper, J. R. (1958). Nuclear migration in the basidiomycete *Schizophyllum commune*. *American Journal of Botany*, 538–546. https://doi.org/10.1002/j.1537-2197.1958.tb13163.x

16. Koltin, Y. and Flexer, A. S. (1969). Alteration of nuclear distribution in B-mutant strains of *Schizophyllum commune*. *Journal of Cell Science*, 4(3), 739–749. https://doi.org/10.1242/jcs.4.3.739

Chapter 10: Clandestine Cooperation

1. Adkar-Purushothama, C. R. and Perreault, J. P. (2020). Current overview on viroid–host interactions. *Wiley Interdisciplinary Reviews: RNA*, 11(2), e1570. https://doi.org/10.1002/wrna.1570

2. Zheludev, I. N. et al. (2024). Viroid-like colonists of human microbiomes. *bioRxiv*. https://doi.org/10.1101/2024.01.20.576352

3. Zwicker, D. et al. (2017). Growth and division of active droplets provides a model for protocells. *Nature Physics*, 13(4), 408–413. https://doi.org/10.1038/nphys3984

4. Dworkin, J. P. et al. (2001). Self-assembling amphiphilic molecules: Synthesis in simulated interstellar/precometary ices. *Proceedings of the National Academy of Sciences*, 98(3), 815–819. https://doi.org/10.1073/pnas.98.3.815

5. Thompson, M. J. and Jiggins, C. (2014). Supergenes and their role in evolution. *Heredity*, 113(1), 1–8. https://doi.org/10.1038/hdy.2014.20

6. Bell, P. J. (2020). Evidence supporting a viral origin of the eukaryotic nucleus. *Virus Research*, 289, 198168. https://doi.org/10.1016/j.virusres.2020.198168

7. Takemura, M. (2020). Medusavirus ancestor in a proto-eukaryotic cell: updating the hypothesis for the viral origin of the nucleus. *Frontiers in Microbiology*, 11, 2169. https://doi.org/10.3389/fmicb.2020.571831

8. Sagan, L. (1967). On the origin of mitosing cells. *Journal of Theoretical Biology*, 14(3), 225-IN6. https://doi.org/10.1016/0022-5193(67)90079-3

9. Martin, W. and Mentel, M. (2010). The origin of mitochondria. *Nature Education*, 3(9), 58.

10. Jensen, P. E. and Leister, D. (2014). Chloroplast evolution, structure and functions. *F1000prime Reports*, 6. https://doi.org/10.12703/P6-40

11. Coale, T. H. et al. (2024). Nitrogen-fixing organelle in a marine alga. *Science*, 384(6692), 217-222. https://doi.org/10.1126/science.adk1075

12. Brunet, T. and King, N. (2017). The origin of animal multicellularity and cell differentiation. *Developmental Cell*, 43(2), 124-140. https://doi.org/10.1016/j.devcel.2017.09.016

13. Matt, G. and Umen, J. (2016). Volvox: A simple algal model for embryogenesis, morphogenesis and cellular differentiation. *Developmental Biology*, 419(1), 99–113. https://doi.org/10.1016/j.ydbio.2016.07.014

14. Olson, L. W. and Kochert, G. (1970). Ultrastructure of *Volvox carteri*: II. The kinetosome. *Archiv für Mikrobiologie*, 74, 31–40. https://doi.org/10.1007/BF00408685

15. Van Donk, E., Ianora, A., and Vos, M. (2011). Induced defences in marine and freshwater phytoplankton: a review. *Hydrobiologia*, 668, 3–19. https://doi.org/10.1007/s10750-010-0395-4

16. Shubert, E., Wilk-Woźniak, E., and Ligęza, S. (2014). An autecological investigation of *Desmodesmus*: implications for ecology and taxonomy. *Plant Ecology and Evolution*, 147(2), 202–212. https://doi.org/10.5091/plecevo.2014.902

17. Cheloni, G. and Slaveykova, V. I. (2021). Morphological plasticity in *Chlamydomonas reinhardtii* and acclimation to micropollutant stress. *Aquatic Toxicology*, 231, 105711. https://doi.org/10.1016/j.aquatox.2020.105711

18. Nelson, D. R. et al. (2024). Macroalgal deep genomics illuminate multiple paths to aquatic, photosynthetic multicellularity. *Molecular Plant*. https://doi.org/10.1016/j.molp.2024.03.011

19. Liu, D. et al. (2023). A cell-based model for size control in the multiple fission alga Chlamydomonas reinhardtii. *Current Biology*, 33(23), 5215–5224. https://doi.org/10.1016/j.cub.2023.10.023

20. Vicencio, J. M. et al. (2008). Senescence, apoptosis or autophagy? When a damaged cell must decide its path–a mini-review. *Gerontology*, 54(2), 92–99. https://doi.org/10.1159/000129697

21. Lowe, S. W. and Lin, A. W. (2000). Apoptosis in cancer. *Carcinogenesis*, 21(3), 485–495. https://doi.org/10.1093/carcin/21.3.485

22. Ryan, F. P. (2004). Human endogenous retroviruses in health and disease: a symbiotic perspective. *Journal of the Royal Society of Medicine*, 97(12), 560–565. https://doi.org/10.1177/014107680409701202

23. Tarlinton, R. E., Meers, J., and Young, P. R. (2006). Retroviral invasion of the koala genome. *Nature*, 442(7098), 79–81. https://doi.org/10.1038/nature04841

24. Voisset, C. et al. (1999). Phylogeny of a novel family of human endogenous retrovirus sequences, HERV-W, in humans and other primates. *AIDS Research and Human Retroviruses*, 15(17), 1529–1533. https://doi.org/10.1089/088922299309810

25. Mi, S. et al. (2000). Syncytin is a captive retroviral envelope protein involved in human placental morphogenesis. *Nature*, 403(6771), 785–789. https://doi.org/10.1038/35001608

26. Wickramage, I. et al. (2023). SINE RNA of the imprinted miRNA clusters mediates constitutive type III interferon expression and antiviral protection in hemochorial placentas. *Cell Host & Microbe*. https://doi.org/10.1016/j.chom.2023.05.018

27. Ghosh, T. et al. (2024). A retroviral link to vertebrate myelination through retrotransposon RNA-mediated control of myelin gene expression. *Cell*, 187(4), 814–830. https://doi.org/10.1016/j.cell.2024.01.011

28. Tillmann, H. L. et al. (2001). Infection with GB virus C and reduced mortality among HIV-infected patients. *New England Journal of Medicine*, 345(10), 715–724. https://doi.org/10.1056/NEJMoa010398

29. Oldstone, M. B. (1988). Prevention of type I diabetes in nonobese diabetic mice by virus infection. *Science*, 239(4839), 500–502. https://doi.org/10.1126/science.239.4839.500

30. Barton, E. S. et al. (2007). Herpesvirus latency confers symbiotic protection from bacterial infection. *Nature*, 447(7142), 326–329. https://doi.org/10.1038/nature05762

31. Marshall, C. W. et al. (2021). Rampant prophage movement among transient competitors drives rapid adaptation during infection. *Science Advances*, 7(29), eabh1489. https://doi.org/10.1126/sciadv.abh1489

32. Johansen, J. et al. (2023). Centenarians have a diverse gut virome with the potential to modulate metabolism and promote healthy lifespan. *Nature Microbiology*, 1–15. https://doi.org/10.1038/s41564-023-01370-6

33. Spragge, F. et al. (2023). Microbiome diversity protects against pathogens by nutrient blocking. *Science*, 382(6676), eadj3502. https://doi.org/10.1126/science.adj3502

34. Lawrence, P. O. (2002). Purification and partial characterization of an entomopoxvirus (DLEPV) from a parasitic wasp of tephritid fruit flies. *Journal of Insect Science*, 2(1), 10. https://doi.org/10.1093/jis/2.1.10

35. Brown, S. P. et al. (2006). Ecology of microbial invasions: amplification allows virus carriers to invade more rapidly when rare. *Current Biology*, 16(20), 2048-2052. https://doi.org/10.1016/j.cub.2006.08.089

36. Fenton, A. et al. (2011). Parasite-induced warning coloration: a novel form of host manipulation. *Animal Behaviour*, 81(2), 417–422. https://doi.org/10.1016/j.anbehav.2010.11.010

37. Asplund, J. and Wardle, D. A. (2017). How lichens impact on terrestrial community and ecosystem properties. *Biological Reviews*, 92(3), 1720–1738. https://doi.org/10.1111/brv.12305

38. Hom, E. F. and Murray, A. W. (2014). Niche engineering demonstrates a latent capacity for fungal–algal mutualism. *Science*, 345(6192), 94–98. https://doi.org/10.1126/science.1253320

39. Spribille, T. et al. (2016). Basidiomycete yeasts in the cortex of ascomycete macrolichens. *Science*, 353(6298), 488–492. https://doi.org/10.1126/science.aaf8287

40. Casano, L. M. et al. (2011). Two *Trebouxia* algae with different physiological performances are ever-present in lichen thalli of *Ramalina farinacea*. Coexistence versus competition? *Environmental Microbiology*, 13(3), 806–818. https://doi.org/10.1111/j.1462-2920.2010.02386.x

41. Wilkinson, D. M. et al. (2015). Are heterotrophic and silica-rich eukaryotic microbes an important part of the lichen symbiosis? *Mycology*, 6(1), 4–7. https://doi.org/10.1080/21501203.2014.974084

42. Petrzik, K. et al. (2019). Chrysoviruses inhabited symbiotic fungi of lichens. *Viruses*, 11(12), 1120. https://doi.org/10.3390/v11121120

Epilogue

1. Jiang, W. and Chen, L. (2021). Alternative splicing: Human disease and quantitative analysis from high-throughput sequencing. *Computational and Structural Biotechnology Journal*, 19, 183–195. https://doi.org/10.1016/j.csbj.2020.12.009

2. Sciarrillo, R. et al. (2020). The role of alternative splicing in cancer: From oncogenesis to drug resistance. *Drug Resistance Updates*, 53, 100728. https://doi.org/10.1016/j.drup.2020.100728

3. Horste, E. L. et al. (2023). Subcytoplasmic location of translation controls protein output. *Molecular Cell*, 83(24), 4509–4523. https://doi.org/10.1016/j.molcel.2023.11.025

4. Murzin, A. G. (2008). Metamorphic proteins. *Science*, 320(5884), 1725–1726. https://doi.org/10.1126/science.1158868

5. Cazaubon, S., Viegas, P., and Couraud, P. O. (2007). Functions of prion protein PrPc. *Medecine Sciences: M/S*, 23(8–9), 741–745. https://doi.org/10.1051/medsci/20072389741

Index

www.ingramcontent.com/pod-product-compliance
Ingram Content Group UK Ltd.
Pitfield, Milton Keynes, MK11 3LW, UK
UKHW040650060725
460485UK00001B/17

9 781784 275778